THE LIFE OF THE SPIDER

BY
J. HENRI FABRE

TRANSLATED BY
ALEXANDER TEIXEIRA DE MATTOS
FELLOW OF THE ZOOLOGICAL SOCIETY OF LONDON

WITH A PREFACE BY MAURICE MAERTERLINCK

ILLUSTRATIONS BY
Clement B. Davis

NEW WEST PRESS

Copyright © 2020 by New West Press

ISBN 978-1-64965-103-7

All rights reserved. This book or any portion thereof may not be reproduced or used in any manner whatsoever without the express written permission of the publisher except for the use of brief quotations in a book review or scholarly journal.

New West Press
Phoenix, AZ 85085
www.nwwst com

Ordering Information:
Special discounts are available on quantity purchases by corporations, associations, educators, and others. For details, contact the publisher at the listed address below.

U.S. trade bookstores and wholesalers: Please contact New West Press:

Tel: (480) 648-1061; or email: contact@nwwst.com

CONTENTS

PREFACE: THE INSECT'S HOMER i

TRANSLATOR'S NOTE XV

I. THE BLACK-BELLIED TARANTULA 1

II. THE BANDED EPEIRA 21

III. THE NARBONNE LYCOSA 35

IV. THE NARBONNE LYCOSA: THE BURROW 46

V. THE NARBONNE LYCOSA: THE FAMILY 61

VI. THE NARBONNE LYCOSA: THE CLIMBING-INSTINCT ... 71

VII. THE SPIDERS' EXODUS 80

VIII. THE CRAB SPIDER 94

IX. THE GARDEN SPIDERS: BUILDING THE WEB 103

X. THE GARDEN SPIDERS: MY NEIGHBOUR 114

XI. THE GARDEN SPIDERS: THE LIME-SNARE 127

XII. THE GARDEN SPIDERS: THE TELEGRAPH-WIRE 133

XIII. THE GARDEN SPIDERS: PAIRING AND HUNTING......... 141

XIV. THE GARDEN SPIDERS: THE QUESTION OF PROPERTY 153

XV. THE LABYRINTH SPIDER 161

XVI. THE CLOTHO SPIDER 177

APPENDIX: THE GEOMETRY OF THE EPEIRA'S WEB 189

PREFACE

THE INSECT'S HOMER

I

ORANGE and Sérignan, the latter a little Provençal village that should be as widely celebrated as Maillane, have of late years rendered honour to a man whose brow deserves to be girt with a double and radiant crown. But fame — at least that which is not the true nor the great fame, but her illegitimate sister, and which creates more noise than durable work in the morning and evening papers — fame is often forgetful, negligent, behindhand or unjust; and the crowd is almost ignorant of the name of J. H. Fabre, who is one of the most profound and inventive scholars and also one of the purest writers and, I was going to add, one of the finest poets of the century that is just past.

J. H. Fabre, as some few people know, is the author of half a score of well-filled volumes in which, under the title of *Souvenirs Entomologiques*, he has set down the results of fifty years of observation, study and experiment on the insects that seem to us the best-known and the most familiar: different species of wasps and wild bees, a few gnats, flies, beetles and caterpillars; in a word, all those vague, unconscious, rudimentary and almost nameless little lives which surround us on every side and which we contemplate with eyes that are amused, but already thinking of other things, when we open our window to welcome the first hours of spring, or when we go into the gardens or the fields to bask in the blue summer days.

---

[1]Maillane is the birthplace of Mistral, the Provençal poet.—*Translator's Note*.

2

We take up at random one of these bulky volumes and naturally expect to find first of all the very learned and rather dry lists of names, the very fastidious and exceedingly quaint specifications of those huge, dusty graveyards of which all the entomological treatises that we have read so far seem almost wholly to consist. We therefore open the book without zest and without unreasonable expectations; and forthwith, from between the open leaves, there rises and unfolds itself, without hesitation, without interruption and almost without remission to the end of the four thousand pages, the most extraordinary of tragic fairy plays that it is possible for the human imagination, not to create or to conceive, but to admit and to acclimatize within itself.

Indeed, there is no question here of the human imagination. The insect does not belong to our world. The other animals, the plants even, notwithstanding their dumb life and the great secrets which they cherish, do not seem wholly foreign to us. In spite of all, we feel a certain earthly brotherhood in them. They often surprise and amaze our intelligence, but do not utterly upset it. There is something, on the other hand, about the insect that does not seem to belong to the habits, the ethics, the psychology of our globe. One would be inclined to say that the insect comes from another planet, more monstrous, more energetic, more insane, more atrocious, more infernal than our own. One would think that it was born of some comet that had lost its course and died demented in space. In vain does it seize upon life with an authority, a fecundity unequalled here below; we cannot accustom ourselves to the idea that it is a thought of that nature of whom we fondly believe ourselves to be the privileged children and probably the ideal to which all the earth's efforts tend. Only the infinitely small disconcerts us still more greatly; but what, in reality, is the infinitely small other than an insect which our eyes do not see? There is, no doubt, in this astonishment and lack of understanding a certain instinctive and profound uneasiness inspired by those existences incomparably better-armed, better-equipped than our own, by those creatures made up of a sort of compressed energy and activity in whom we suspect our most mysterious adversaries, our ultimate rivals and, perhaps, our successors.

3

But it is time, under the conduct of an admirable guide, to penetrate behind the scenes of our fairy play and to study at close quarters the actors and supernumeraries, loathsome or magnificent, as the case may be, grotesque or sinister, heroic or appalling, genial or stupid and almost always improbable and unintelligible.

And here, to begin with, taking the first that comes, is one of those individuals, frequent in the South, where we can see it prowling around the abundant manna which the mule scatters heedlessly along the white roads and the stony paths: I mean the Sacred Scarab of the Egyptians, or, more simply, the Dung-beetle, the brother of our northern Geotrupes, a big Coleopteron all clad in black, whose mission in this world is to shape the more savoury parts of the prize into an enormous ball which he must next roll to the subterranean dining-room where the incredible digestive adventure is to take its course. But destiny, jealous of all undiluted bliss, before admitting him to that spot of sheer delight, imposes upon the grave and probably sententious beetle tribulations without number, which are nearly always complicated by the arrival of an untoward parasite.

Hardly has he begun, by dint of great efforts of his frontal shield and bandy legs, to roll the toothsome sphere backwards, when an indelicate colleague, who has been awaiting the completion of the work, appears and hypocritically offers his services. The other well knows that, in this case, help and services, besides being quite unnecessary, will soon mean partition and dispossession; and he accepts the enforced collaboration without enthusiasm. But, so that their respective rights may be clearly marked, the legal owner invariably retains his original place, that is to say, he pushes the ball with his forehead, whereas the compulsory guest, on the other side, pulls it towards him. And thus it jogs along between the two gossips, amid interminable vicissitudes, flurried falls, grotesque tumbles, till it reaches the place chosen to receive the treasure and to become the banqueting-hall. On arriving, the owner sets about digging out the refectory, while the sponger pretends to go innocently to sleep on the top of the bolus. The excavation becomes visibly wider and deeper; and soon the first dung-beetle dives bodily into it. This is the moment for which the cunning auxiliary was waiting. He nimbly scrambles down from the blissful eminence and, pushing it with all the energy that a bad conscience gives, strives to gain the offing. But the other, who is rather distrustful, interrupts his laborious excavations, looks over-board, sees the sacrilegious rape and leaps out of the hole. Caught in the act, the shame-less and dishonest partner makes untold efforts to play upon the other's credulity, turns round and round the inestimable orb and, embracing it and propping himself against it, with fraudulent heroic exertions pretends to be frantically supporting it on a non-existent slope. The two expostulate with each other in silence, gesticulate wildly with their mandibles and tarsi and then, with one accord, bring back the ball to the burrow.

It is pronounced sufficiently spacious and comfortable. They introduce the treasure, they close the entrance to the corridor; and now, in the propitious darkness and the warm damp, where the magnificent stercoral globe alone holds sway, the two reconciled mess- mates sit down face to face. Then, far from the light and the cares of day and in the great silence of the hypogeous shade, solemnly commences the most fabulous ban- quet whereof abdominal imagination ever evoked the absolute beatitudes.

For two whole months, they remain cloistered; and, with their paunches proportionately hollowing out the inexhaustible sphere, definite archetypes and sovereign symbols of the pleasures of the table and the gaiety of the belly, they eat without stopping, without Interrupting themselves for a second, day or night. And, while they gorge, steadily, with a movement perceptible and constant as that of a clock, at the rate of three millimetres a minute, an endless, unbroken ribbon unwinds and stretches Itself behind them, fixing the memory and recording the hours, days and weeks of the prodigious feast.

4

After the Dung-beetle, that dolt of the company, let us greet, also in the order of the Coleoptera, the model household of the *Minotaurus typhæis*, which is pretty well-known and extremely gentle. In spite of its dreadful name. The female digs a huge burrow which is often more than a yard and a half deep and which consists of spiral staircases, landings, passages and numerous chambers. The male loads the earth on the three-pronged fork that surmounts his head and carries it to the entrance of the conjugal dwelling. Next, he goes into the fields in search of the harmless droppings left by the sheep, takes them down to the first storey of the crypt and reduces them to flour with his trident, while the mother, right at the bottom, collects the flour and kneads it into huge cylindrical loaves, which will presently be food for the little ones. For three months, until the provisions are deemed sufficient, the unfortunate husband, without taking nourishment of any kind, exhausts himself in this gigantic work. At last, his task accomplished, feeling his end at hand, so as not to encumber the house with his wretched remains, he spends his last strength in leaving the burrow, drags himself laboriously along and, lonely and resigned, knowing that he is henceforth good for nothing, goes and dies tar away among the stones.

Here, on another side, are some rather strange caterpillars, the Proressionaries, which are not rare; and, as it happens, a single string of them, five or six yards long, has just climbed down from my umbrella-pines and is at this moment unfolding Itself in the walks of my garden, carpeting the ground traversed with transparent silk, according to the custom of the race. To say nothing of the meteorological apparatus of unparalleled delicacy which they carry on their backs, these caterpillars, as everybody knows, have this remarkable quality, that they travel only in a troop, one after the other, like Breughel's blind men or those of the parable, each of them obstinately, indissolubly following its leader; so much so that, our author having one morning disposed the file on the edge of a large stone vase, thus closing the circuit, for seven whole days, during an atrocious week, amidst cold, hunger and unspeakable weariness, the unhappy troop on its tragic round, without rest, respite or mercy, pursued the pitiless circle until death overtook it.

5

But I see that our heroes are infinitely too numerous and that we must not linger over our descriptions. We may at most, in enumerating the more important and familiar, bestow on each of them a hurried epithet, in the manner of old Homer. Shall I mention, for instance, the Leucospis, a parasite of the Mason-bee, who, to slay his brothers and sisters in their cradle, arms himself with a horn helmet and a barbed breastplate, which he doffs immediately after the extermination, the safeguard of a hideous right of primogeniture? Shall I tell of the marvellous anatomical knowledge of the Tachytes, of the Cerceris, of the Ammophila, of the Languedocian Sphex, who, according as they wish to paralyze or to kill their prey or their adversary, know exactly, without ever blundering, which nerve-centre to strike with their sting or their mandibles? Shall I speak of the art of the Eumenes, who transforms her stronghold into a complete museum adorned with shells and grains of translucent quartz; of the magnificent metamorphosis of the *Pachytilus cinarescens*; of the musical instrument owned by the Cricket, whose bow numbers one hundred and fifty triangular prisms that set in motion simultaneously the four dulcimers of the elytron? Shall I sing the fairy-like birth of the nymphs of the Anthophagus, a transparent monster, with a bull's snout, that seems carved out of a block of crystal? Would you behold the Flesh-fly, the common Blue-bottle, daughter of the maggot, as she issues from the earth? Listen to our author:

'She disjoints her head into two movable halves, which, each distended with its great red eye, by turns separate and reunite. In the intervening space a large glassy hernia rises and disappears, disappears and rises. When the two halves move asunder, with one eye forced back to the right and the other to the left, it is as though the insect were splitting its brain-pan in order to expel the contents. Then the hernia rises, blunt at the end and swollen into a great knob. Next, the forehead closes and the hernia retreats, leaving visible only a kind of shapeless muzzle. In short, a frontal pouch, with deep pulsations momentarily renewed, becomes the instrument of deliverance, the pestle wherewith the newly-hatched Dipteron bruises the sand and causes it to crumble. Gradually, the legs push the rubbish back and the insect advances so much towards the surface.'

6

And monster after monster passes, such as the imagination of Bosch or Callot never conceived! The larva of the Rose-chafer, which, though it have legs under its belly, always travels on its back;

the Blue-winged Locust, unluckier still than the Flesh-fly and possessing nothing wherewith to perforate the soil, to escape from the tomb and reach the light but a cervical bladder, a viscous blister; and the Empusa, who, with her curved abdomen, her great projecting eyes, her legs with knee-pieces armed with cleavers, her halberd, her abnormally tall mitre would certainly be the most devilish goblin that ever walked the earth, if, beside her, the Praying Mantis were not so frightful that her mere aspect deprives her victims of their power of movement when she assumes, in front of them, what the entomologists have termed 'the spectral attitude.'

One cannot mention, even casually, the numberless Industries — nearly all of absorbing interest — exercised among the rocks, under the ground, in the walls, on the branches, the grass, the flowers, the fruits and down to the very bodies of the subjects studied; for we sometimes find a treble superposition of parasites, as in the Oil-beetles ; and we see the maggot itself, the sinister guest at the last feast of all, feed some thirty brigands with its substance.

7

Among the Hymenoptera, which represent the most intellectual class in the world which we are studying, the building-talents of our wonderful Domestic Bee are certainly equal, in other orders of architecture, by those of more than one wild and solitary bee and notably by the Megachile, or Leaf-cutter, a little insect which is not all outside show and which, to house its eggs, manufactures honey-pots formed of a multitude of disks and ellipses cut with mathematical precision from the leaves of certain trees. For lack of space, I am unable, to my great regret, to quote the beautiful and pellucid pages which J. H. Fabre, with his usual conscientiousness, devotes to the exhaustive study of this admirable work; nevertheless, since the occasion offers, let us listen to his own words, though it be but for a moment and in regard to a single detail:

'With the oval pieces, the question changes. What model has the Megachile when cutting into fine ellipses the delicate material of the robinia? What ideal pattern guides her scissors? What measure dictates the dimensions? One would like to think of the insect as a living compass, capable of tracing an elliptic curve by a certain natural inflexion of the body, even as our arm traces a circle by swinging from the shoulder. A blind mechanism, the mere outcome of her organization, would in that case be responsible for her geometry. This explanation would tempt me, if the oval pieces of large dimensions were not accompanied by much smaller, but likewise oval pieces, to fill the empty spaces. A compass which changes its radius of itself and alters the degree of curvature according to the exigencies of a plan appears to me an instrument somewhat difficult to believe in. There must be something better than that. The circular pieces of the lid suggest it to us.

'If, by the mere flexion inherent in her structure, the leaf-cutter succeeds in cutting out ovals, how does she manage to cut out rounds? Can we admit the presence of other wheels in the machinery for the new pattern, so different in shape and size? However, the real point of the difficulty does not lie there. Those rounds, for the most part, fit the mouth of the bottle with almost exact precision. When the cell is finished, the bee flies hundreds of yards further to make the lid. She arrives at the leaf from which the disk is to be cut. What picture, what recollection has she of the pot to be covered? Why, none at all: she has never seen it; she works underground, in profound darkness! At the utmost, she can have the indications of touch: not actual indications, of course, for the pot is not there, but past indications, ineffective in a work of precision. And yet the disk must be of a fixed diameter: if it were too large, it would not fit in; if too small, it would close badly, it would smother the egg by sliding down on the honey. How shall it be given its correct dimensions without a pattern? The Bee does not hesitate for a moment. She cuts out her disk with the same rapidity which she would display in detaching any shapeless lobe just useful for closing; and that disk, without further measurement, is of the right size to fit the pot. Let whoso will explain this geometry, which in my opinion is inexplicable, even when we allow for memory begotten of touch and sight.'

Let us add that the author has calculated that, to form the cells of a kindred Mega- chile, the Silky Megachile, exactly 1,064 of these ellipses and disks would be required; and they must all be collected and shaped in the course of an existence that lasts a few weeks.

8

Who would imagine that the Pentatomida, on the other hand, the poor and evil-smelling bug of the woods, has invented a really extraordinary apparatus wherewith to leave the egg? And first let us state that this egg is a marvellous little box of snowy whiteness, which our author thus describes:

'The microscope discovers a surface engraved with dents similar to those of a thimble and arranged with exquisite symmetry. At the top and bottom of the cylinder is a wide belt of a dead black; on the sides, a large white zone with four big, black spots evenly distributed. The lid, surrounded by snowy cilia and encircled with white at the edge, swells into a black cap with a white knot in the centre. Altogether, a dismal burial urn, with the sudden contrast between the dead black and the fleecy white. The funeral pottery of the ancient Etruscans would have found a magnificent model here.'

The little bug, whose forehead is too soft, covers her head, to raise the lid of the box, with a mitre formed of three triangular rods, which is always at the bottom of the egg at the moment of delivery. Her limbs being sheathed like those of a mummy, she has nothing wherewith to put her tringles in motion except the pulsations produced by the rhythmic flow of blood in her skull and acting after the manner of a piston. The rivets of the lid gradually give way; and, as soon as the insect is free, she lays aside her mechanical helmet.

Another species of bug, the *Reduvius personatus*, which lives mostly in lumber-rooms, where it lies hidden in the dust, has invented a still more astonishing system of hatching. Here, the lid of the egg is not riveted, as in the case of the Pentatomidæ, but simply glued. At the moment of liberation, the lid rises and we see:

'...a spherical vesicle emerge from the shell and gradually expand, like a soap-bubble blown through a straw. Driven further and further back by the extension of this bladder, the lid falls.

'Then the bomb bursts; in other words, the blister, swollen beyond its capacity of resistance, rips at the top. This envelope, which is an extremely tenuous membrane, generally remains clinging to the edge of the orifice, where it forms a high, white rim. At other times, the explosion loosens it and flings it outside the shell. In those conditions, it is a dainty cup, half spherical, with torn edges, lengthened out below into a delicate, winding stalk.'

Now, how is this miraculous explosion produced? J. H. Fabre assumes that:

'Very slowly, as the little animal takes shape and grows, this bladder-shaped reservoir receives the products of the work of respiration performed under the cover of the outer membrane. Instead of being expelled through the egg-shell, the carbonic acid, the incessant result of the vital oxidization, is accumulated in this sort of gasometer, inflates and distends it and presses upon the lid. When the insect is ripe for hatching, a super-added activity in the respiration completes the inflation, which perhaps has been preparing since the first evolution of the germ. At last, yielding to the increasing pressure of the gaseous bladder, the lid becomes unsealed. The Chick in its shell has its air-chamber; the young Reduvius has its bomb of carbonic acid: it frees itself in the act of breathing.'

One would never weary of dipping eagerly into these inexhaustible treasures. We imagine, for instance, that, from seeing cobwebs so frequently displayed in all manner of places, we possess adequate notions of the genius and methods of our familiar spiders. Far from it: the realities of scientific observation call for an entire volume crammed with revelations of which we had no conception. I will simply name, at random, the symmetrical arches of the Clotho Spider's nest, the astonishing funicular flight of the young of our Garden Spider, the diving-bell of the Water Spider, the live telephone-wire which connects the web with the leg of the Cross Spider hidden in her parlour and informs her whether the vibration of her toils is due to the capture of a prey or a caprice of the wind.

9

It is impossible, therefore, short of having unlimited space at one's disposal, to do more than touch, as it were with the tip of the phrases, upon, the miracles of maternal instinct, which, moreover, are confounded with those of the higher manufactures and form the bright centre of the insect's psychology. One would, in the same way, require several chapters to convey a summary idea of the nuptial rites which constitute the quaintest and most fabulous episodes of these new Arabian Nights.

The male of the Spanish-fly, for instance, begins by frenziedly beating his spouse with his abdomen and his feet, after which, with his arms crossed and quivering, he remains long in ecstasy. The newly-wedded Osmiæ clap their mandibles terribly, as though it were a matter rather of devouring each other; on the other hand, the largest of our moths, the Great Peacock, who is the size of a bat, when drunk with love finds his mouth so completely atrophied that it becomes no more than a vague shadow. But nothing equals the marriage of the Green Grasshopper, of which I cannot speak here, for it is doubtful whether even the Latin language possesses the words needed to describe it as it should be described.

All said, the marriage customs are dreadful and, contrary to that which happens in every other world, here it is the female of the pair that stands for strength and intelligence and also for cruelty and tyranny, which appear to be their inevitable consequence. Almost every wedding ends in the violent and immediate death of the husband. Often, the bride begins by eating a certain number of suitors. The archetype of these fantastic unions could be supplied by the Languedocian Scorpions, who, as we know, carry lobster-claws and a long tail supplied with a sting, the prick of which is extremely dangerous. They have a prelude to the festival in the shape of a sentimental stroll, claw in claw; then, motionless, with fingers still gripped, they contemplate each other blissfully, interminably: day and night pass over their ecstasy while they remain face to face, petrified with admiration. Next, the foreheads come together and touch; the mouths — if we can give the name of mouth to the monstrous orifice that opens between the claws — are joined in a sort of kiss; after which the union is accomplished, the male is transfixed with a mortal sting and the terrible spouse crunches and gobbles him up with gusto.

But the Mantis, the ecstatic insect with the arms always raised in an attitude of supreme invocation, the horrible *Mantis religiosa* or Praying Mantis, does better still: she eats her husbands (for the insatiable creature sometimes consumes seven or eight in succession), while they strain her passionately to their heart. Her inconceivable kisses devour, not metaphorically, but in an appallingly real fashion, the ill-fated choice of her soul or her stomach. She begins with the head, goes down to the thorax, nor stops till she comes to the hind-legs, which she deems too tough. She then pushes away the unfortunate remains, while a new lover, who was quietly awaiting the end of the monstrous banquet, heroically steps forward to undergo the same fate.

J. H. Fabre is indeed the revealer of this new world, for, strange as the admission may seem at a time when we think that we know all that surrounds us, most of those insects minutely described in the vocabularies, learnedly classified and barbarously christened had hardly ever been observed in real life or thoroughly investigated, in all the phases of their brief and evasive appearances. He has devoted to surprising their little secrets, which are the reverse of our greatest mysteries, fifty years of a solitary existence, misunderstood, poor, often very near to penury, but lit up every day by the joy which a truth brings, which is the greatest of all human joys. Petty truths, I shall be told, those presented by the habits of a spider or a grasshopper. There are no petty truths to-day; there is but one truth, whose looking-glass, to our uncertain eyes, seems broken, though its every fragment, whether reflecting the evolution of a planet or the flight of a bee, contains the supreme law.

And these truths thus discovered had the good fortune to be grasped by a mind which knew how to understand what they themselves can but ambiguously express, to interpret what they are obliged to conceal and, at the same time, to appreciate the shimmering beauty, almost invisible to the majority of mankind, that shines for a moment around all that exists, especially around that which still remains very close to nature and has hardly left its primeval obscurity.

To make of these long annals the generous and delightful masterpiece that they are and not the monotonous and arid register of little descriptions and insignificant acts that they might have been, various and so to speak conflicting gifts were needed. To the patience, the precision, the scientific minuteness, the protean and practical ingenuity, the energy of a Darwin in the face of the unknown, to the faculty of expressing what has to be expressed with order, clearness and certainty, the venerable anchorite of Sèrignan adds many of those qualities which are not to be acquired, certain of those innate good poetic virtues which cause his sure and supple prose, devoid of artificial ornament and yet adorned with simple and as it were unintentional charm, to take its place among the excellent and lasting prose of the day, prose of the kind that has its own atmosphere, in which we breathe gratefully and tranquilly and which we find only around masterpieces.

Lastly, there was needed — and this was not the least requirement of the work — a mind ever ready to cope with the riddles which, among those little objects, rise up at every step, as enormous as those which fill the skies and perhaps more numerous, more imperious and more strange, as though nature had here given a freer scope to her last wishes and an easier outlet to her secret thoughts. He shrinks from none of those boundless problems which are persistently put to us by all the inhabitants of that tiny world where mysteries are heaped up in a denser and more bewildering fashion than in any other. He thus meets and faces, turn by turn, the redoubtable questions of instinct and intelligence, of the origin of species, of the harmony or the accidents of the universe, of the life lavished upon the abysses of death without counting the no less vast, but so to speak more human problems which, among infinite others, are inscribed within the range, if not within the grasp, of our intelligence: parthenogenesis; the prodigious geometry of the wasps and bees; the logarithmic spiral of the Snail; the antennary sense; the miraculous force which, in absolute isolation, without the possible introduction of anything from the outside, increases the volume of the Minotaurus' egg tenfold, where it lies, and, during seven to nine months, nourishes with an invisible and spiritual food, not the lethargy, but the active life of the Scorpion and of the young of the Lycosa and the Clotho Spider. He does not attempt to explain them by one of those generally-acceptable theories such as that of evolution, which merely shifts the ground of the difficulty and which, I may mention in passing, emerges from these volumes in a somewhat sorry plight, after being sharply confronted with incontestable facts.

Waiting for chance or a god to enlighten us, he is able, in the presence of the unknown, to preserve that great religious and attentive silence which is dominant in the best minds of the day. There are those who say:

'Now that you have reaped a plentiful harvest of details, you should follow up analysis with synthesis and generalize the origin of instinct in an all-embracing view.'

To these he replies, with the humble and magnificent loyalty that illumines all his work:

'Because I have stirred a few grains of sand on the shore, am I in a position to know the depths of the ocean?

'Life has unfathomable secrets. Human knowledge will be erased from the archives of the world before we possess the last word that the Gnat has to say to us....

'Success is for the loud talkers, the self-convinced dogmatists; everything is admitted on condition that it be noisily proclaimed. Let us throw off this sham and recognize that, in reality, we know nothing about anything, if things were probed to the bottom. Scientifically, Nature is a riddle without a definite solution to satisfy man's curiosity. Hypothesis follows on hypothesis; the theoretical rubbish-heap accumulates; and truth ever eludes us. To know how not to know might well be the last word of wisdom.'

Evidently, this is hoping too little. In the frightful pit, in the bottomless funnel wherein whirl all those contradictory facts which are resolved in obscurity, we know just as much as our cave-dwelling ancestors; but at least we know that we do not know. We survey the dark faces of all the riddles, we try to estimate their number, to classify their varying degrees of dimness, to obtain an idea of their places and extent. That already is something, pending the day of the first gleams of light. In any case, it means doing, in the presence of the mysteries, all that the most upright intelligence can do to-day; and that is what the author of this incomparable Iliad does, with more confidence than he professes. He gazes at them attentively. He wears out his life in surprising their most minute secrets. He prepares for them, in his thoughts and in ours, the field necessary for their evolutions. He increases the consciousness of his ignorance in proportion to their importance and learns to understand more and more that they are incomprehensible.

MAURICE MAETERLINICK

TRANSLATOR'S NOTE

The following essays have been selected from the ten volumes composing the *Souvenirs entomologiques*. Although a good deal of Henri Fabre's masterpiece has been published in English, none of the articles treating of spiders has been issued before, with the exception of that forming Chapter II of the present volume. *The Banded Epeira*, which first appeared in *The English Review*. The rest are new to England and America.

The Fabre books already published are *Insect Life*, translated by the author of Mademoiselle Mori (Macmillan Co., 1901); *The Life and Love of the Insect*, translated by myself (Macmillan Co., 1911) and *Social Life in the Insect World*, translated by Mr. Bernard Miall (Century Co., 1912). References to the above volumes will be found, whenever necessary, in the foot-notes to the present edition.

For the rest, I have tried not to overburden my version with notes; and, in view of this, I have, as far as possible, simplified the scientific terms that occur in the text. In so doing I know that I have but followed the wishes of the author, who never wearies of protesting against 'the barbarous terminology' favoured by his brother-naturalists. The matter became even more urgent in English than in any of the Latin languages; and I readily agreed when it was pointed out to me that, in a work essentially intended for general reading, there was no purpose in speaking of a Coleopteron when the word 'beetle' was to hand. In cases where an insect had inevitably to be mentioned by its Greek or Latin name, a note is given explaining, in the fewest words, the nature of the insect in question.

I have to thank my friend, M. Maurice Maeterlinck, for the stately preface which he has contributed to this volume, and Mr. Marmaduke Langdale and Miss Frances Rodwell for the generous assistance which they have given me in the details of my work. And I am also greatly indebted to Mr. W. S. Graff Baker for his invaluable help with the mathematical difficulties that confronted me in the translation of the Appendix.

ALEXANDER TEIXEIRA DE MATTOS.
CHELSEA, 10 *October*, 1912.

CHAPTER I

THE BLACK-BELLIED TARANTULA

THE Spider has a bad name: to most of us, she represents an odious, noxious animal, which every one hastens to crush under foot. Against this summary verdict the observer sets the beast's industry, its talent as a weaver, its wiliness in the chase, its tragic nuptials and other characteristics of great interest. Yes, the Spider is well worth studying, apart from any scientific reasons; but she is said to be poisonous and that is her crime and the primary cause of the repugnance wherewith she inspires us. Poisonous, I agree, if by that we understand that the animal is armed with two fangs which cause the immediate death of the little victims which it catches; but there is a wide difference between killing a Midge and harming a man. However immediate in its effects upon the insect entangled in the fatal web, the Spider's poison is not serious for us and causes less inconvenience than a Gnat-bite. That, at least, is what we can safely say as regards the great majority of the Spiders of our regions.

Nevertheless, a few are to be feared; and foremost among these is the Malmignatte, the terror of the Corsican peasantry. I have seen her settle in the furrows, lay out her web and rush boldly at insects larger than herself; I have admired her garb of black velvet speckled with carmine-red; above all, I have heard most disquieting stories told about her. Around Ajaccio and Bonifacio, her bite is reputed very dangerous, sometimes mortal. The countryman declares this for a fact and the doctor does not always dare deny it. In the neighbourhood of Pujaud, not far from Avignon, the harvesters speak with dread of *Theridion lugubre*,[1] first observed by Léon Dufour in the Catalonian mountains; according to them, her bite

---

[1] A small or moderate-sized spider found among foliage.—*Translator's Note*.

would lead to serious accidents. The Italians have bestowed a bad reputation on the Tarantula, who produces convulsions and frenzied dances in the person stung by her. To cope with 'tarantism,' the name given to the disease that follows on the bite of the Italian Spider, you must have recourse to music, the only efficacious remedy, so they tell us. Special tunes have been noted, those quickest to afford relief. There is medical choreography, medical music. And have we not the tarentella, a lively and nimble dance, bequeathed to us perhaps by the healing art of the Calabrian peasant?

Must we take these queer things seriously or laugh at them? From the little that I have seen, I hesitate to pronounce an opinion. Nothing tells us that the bite of the Tarantula may not provoke, in weak and very impressionable people, a nervous disorder which music will relieve; nothing tells us that a profuse perspiration, resulting from a very energetic dance, is not likely to diminish the discomfort by diminishing the cause of the ailment. So far from laughing, I reflect and enquire, when the Calabrian peasant talks to me of his Tarantula, the Pujaud reaper of his *Theridion lugubre*, the Corsican husbandman of his Malmignatte. Those Spiders might easily deserve, at least partly, their terrible reputation.

The most powerful Spider in my district, the Black-bellied Tarantula, will presently give us something to think about, in this connection. It is not my business to discuss a medical point, I interest myself especially in matters of instinct; but, as the poison-fangs play a leading part in the huntress' manoeuvres of war, I shall speak of their effects by the way. The habits of the Tarantula, her ambushes, her artifices, her methods of killing her prey: these constitute my subject. I will preface it with an account by Léon Dufour,[1] one of those accounts in which I used to delight and which did much to bring me into closer touch with the insect. The Wizard of the Landes tells us of the ordinary Tarantula, that of the Calabrias, observed by him in Spain:

'*Lycosa tarantula* by preference inhabits open places, dry, arid, uncultivated places, exposed to the sun. She lives generally—at least when full-grown—in underground passages, regular burrows,

---

[1] Léon Dufour (1780-1865) was an army surgeon who served with distinction in several campaigns and subsequently practised as a doctor in the Landes. He attained great eminence as a naturalist.—*Translator's Note*.

which she digs for herself. These burrows are cylindrical; they are often an inch in diameter and run into the ground to a depth of more than a foot; but they are not perpendicular. The inhabitant of this gut proves that she is at the same time a skilful hunter and an able engineer. It was a question for her not only of constructing a deep retreat that could hide her from the pursuit of her foes: she also had to set up her observatory whence to watch for her prey and dart out upon it. The Tarantula provides for every contingency: the underground passage, in fact, begins by being vertical, but, at four or five inches from the surface, it bends at an obtuse angle, forms a horizontal turning and then becomes perpendicular once more. It is at the elbow of this tunnel that the Tarantula posts herself as a vigilant sentry and does not for a moment lose sight of the door of her dwelling; it was there that, at the period when I was hunting her, I used to see those eyes gleaming like diamonds, bright as a cat's eyes in the dark.

'The outer orifice of the Tarantula's burrow is usually surmounted by a shaft constructed throughout by herself. It is a genuine work of architecture, standing as much as an inch above the ground and sometimes two inches in diameter, so that it is wider than the burrow itself. This last circumstance, which seems to have been calculated by the industrious Spider, lends itself admirably to the necessary extension of the legs at the moment when the prey is to be seized. The shaft is composed mainly of bits of dry wood joined by a little clay and so artistically laid, one above the other, that they form the scaffolding of a straight column, the inside of which is a hollow cylinder. The solidity of this tubular building, of this outwork, is ensured above all by the fact that it is lined, upholstered within, with a texture woven by the Lycosa's[1] spinnerets and continued throughout the interior of the burrow. It is easy to imagine how useful this cleverly-manufactured lining must be for preventing landslip or warping, for maintaining cleanliness and for helping her claws to scale the fortress.

'I hinted that this outwork of the burrow was not there invariably; as a matter of fact, I have often come across Tarantulas'

---

[1]The Tarantula is a Lycosa, or Wolf-spider. Fabre's Tarantula, the Black-bellied Tarantula, is identical with the Narbonne Lycosa, under which name the description is continued in Chapters iii. to vi., all of which were written at a considerably later date than the present chapter.—*Translator's Note*.

holes without a trace of it, perhaps because it had been accidentally destroyed by the weather, or because the Lycosa may not always light upon the proper building-materials, or, lastly, because architectural talent is possibly declared only in individuals that have reached the final stage, the period of perfection of their physical and intellectual development.

'One thing is certain, that I have had numerous opportunities of seeing these shafts, these out-works of the Tarantula's abode; they remind me, on a larger scale, of the tubes of certain Caddis-worms. The Arachnid had more than one object in view in constructing them: she shelters her retreat from the floods; she protects it from the fall of foreign bodies which, swept by the wind, might end by obstructing it; lastly, she uses it as a snare by offering the Flies and other insects whereon she feeds a projecting point to settle on. Who shall tell us all the wiles employed by this clever and daring huntress?

'Let us now say something about my rather diverting Tarantula-hunts. The best season for them is the months of May and June. The first time that I lighted on this Spider's burrows and discovered that they were inhabited by seeing her come to a point on the first floor of her dwelling—the elbow which I have mentioned—I thought that I must attack her by main force and pursue her relentlessly in order to capture her; I spent whole hours in opening up the trench with a knife a foot long by two inches wide, without meeting the Tarantula. I renewed the operation in other burrows, always with the same want of success; I really wanted a pickaxe to achieve my object, but I was too far from any kind of house. I was obliged to change my plan of attack and I resorted to craft. Necessity, they say, is the mother of invention.

'It occurred to me to take a stalk, topped with its spikelet, by way of a bait, and to rub and move it gently at the orifice of the burrow. I soon saw that the Lycosa's attention and desires were roused. Attracted by the bait, she came with measured steps towards the spikelet. I withdrew it in good time a little outside the hole, so as not to leave the animal time for reflexion; and the Spider suddenly, with a rush, darted out of her dwelling, of which I hastened to close the entrance. The Tarantula, bewildered by her unaccustomed liberty, was very awkward in evading my attempts at capture; and I compelled her to enter a paper bag, which I closed without delay.

'Sometimes, suspecting the trap, or perhaps less pressed by hunger, she would remain coy and motionless, at a slight distance from the threshold, which she did not think it opportune to cross. Her patience outlasted mine. In that case, I employed the following tactics: after making sure of the Lycosa's position and the direction of the tunnel, I drove a knife into it on the slant, so as to take the animal in the rear and cut off its retreat by stopping up the burrow. I seldom failed in my attempt, especially in soil that was not stony. In these critical circumstances, either the Tarantula took fright and deserted her lair for the open, or else she stubbornly remained with her back to the blade. I would then give a sudden jerk to the knife, which flung both the earth and the Lycosa to a distance, enabling me to capture her. By employing this hunting-method, I sometimes caught as many as fifteen Tarantulae within the space of an hour.

'In a few cases, in which the Tarantula was under no misapprehension as to the trap which I was setting for her, I was not a little surprised, when I pushed the stalk far enough down to twist it round her hiding-place, to see her play with the spikelet more or less contemptuously and push it away with her legs, without troubling to retreat to the back of her lair.

'The Apulian peasants, according to Baglivi's[1] account, also hunt the Tarantula by imitating the humming of an insect with an oat-stalk at the entrance to her burrow. I quote the passage:

'"*Ruricolæ nostri quando eas captare volunt, ad illorum latibula accedunt, tenuisque avenacæ fistulæ sonum, apum murmuri non absimilem, modulantur. Quo audito, ferox exit Tarentula ut muscas vel alia hujus modi insecta, quorum murmur esse putat, captat; captatur tamen ista a rustico insidiatore.*"[2]

'The Tarantula, so dreadful at first sight, especially when we are filled with the idea that her bite is dangerous, so fierce in appearance, is nevertheless quite easy to tame, as I have often found by experiment.

---

[1] Giorgio Baglivi (1669-1707), professor of anatomy and medicine at Rome.—*Translator's Note.*

[2] 'When our husbandmen wish to catch them, they approach their hiding-places, and play on a thin grass pipe, making a sound not unlike the humming of bees. Hearing which, the Tarantula rushes out fiercely that she may catch the flies or other insects of this kind, whose buzzing she thinks it to be; but she herself is caught by her rustic trapper.'

'On the 7th of May 1812, while at Valencia, in Spain, I caught a fair-sized male Tarantula, without hurting him, and imprisoned him in a glass jar, with a paper cover in which I cut a trap-door. At the bottom of the jar I put a paper bag, to serve as his habitual residence. I placed the jar on a table in my bedroom, so as to have him under frequent observation. He soon grew accustomed to captivity and ended by becoming so familiar that he would come and take from my fingers the live Fly which I gave him. After killing his victim with the fangs of his mandibles, he was not satisfied, like most Spiders, to suck her head: he chewed her whole body, shoving it piecemeal into his mouth with his palpi, after which he threw up the masticated teguments and swept them away from his lodging.

'Having finished his meal, he nearly always made his toilet, which consisted in brushing his palpi and mandibles, both inside and out, with his front tarsi. After that, he resumed his air of motionless gravity. The evening and the night were his time for taking his walks abroad. I often heard him scratching the paper of the bag. These habits confirm the opinion, which I have already expressed elsewhere, that most Spiders have the faculty of seeing by day and night, like cats.

'On the 28th of June, my Tarantula cast his skin. It was his last moult and did not perceptibly alter either the colour of his attire or the dimensions of his body. On the 14th of July, I had to leave Valencia; and I stayed away until the 23rd. During this time, the Tarantula fasted; I found him looking quite well on my return. On the 20th of August, I again left for a nine days' absence, which my prisoner bore without food and without detriment to his health. On the 1st of October, I once more deserted the Tarantula, leaving him without provisions. On the 21st, I was fifty miles from Valencia and, as I intended to remain there, I sent a servant to fetch him. I was sorry to learn that he was not found in the jar, and I never heard what became of him.

'I will end my observations on the Tarantulae with a short description of a curious fight between those animals. One day, when I had had a successful hunt after these Lycosae, I picked out two full-grown and very powerful males and brought them together in a wide jar, in order to enjoy the sight of a combat to the death. After walking round the arena several times, to try and avoid each other, they were not slow in placing themselves in a warlike attitude, as though at a given signal. I saw them, to my surprise, take their distances and sit up solemnly on their hind-legs, so as mutually to present the shield of their chests to each other. After watching them face to face like that for two minutes, during which they had doubtless provoked each other by glances that escaped my own, I saw them fling themselves upon each other at the same time, twisting their legs round each other and obstinately struggling to bite each other with the fangs of the mandibles. Whether from fatigue or from convention, the combat was suspended; there was a few seconds' truce; and each athlete moved away and resumed his threatening posture. This circumstance reminded me that, in the strange fights between cats, there are also suspensions of hostilities. But the contest was soon renewed between my two Tarantulae with increased fierceness. One of them, after holding victory in the balance for a while, was at last thrown and received a mortal wound in the head. He became the prey of the conqueror, who tore open his skull and devoured it. After this curious duel, I kept the victorious Tarantula alive for several weeks.'

My district does not boast the ordinary Tarantula, the Spider whose habits have been described above by the Wizard of the Landes; but it possesses an equivalent in the shape of the Black-bellied Tarantula, or Narbonne Lycosa, half the size of the other, clad in black velvet on the lower surface, especially under the belly, with brown chevrons on the abdomen and grey and white rings around the legs. Her favourite home is the dry, pebbly ground, covered with sun-scorched thyme. In my *harmas*[1] laboratory there are quite twenty of this Spider's burrows. Rarely do I pass by one of these haunts without giving a glance down the pit where gleam, like diamonds, the four great eyes, the four telescopes, of the hermit. The four others, which are much smaller, are not visible at that depth.

Would I have greater riches, I have but to walk a hundred yards from my house, on the neighbouring plateau, once a shady forest, to-day a dreary solitude where the Cricket browses and the Wheat-ear flits from stone to stone. The love of lucre has laid waste the

---

[1] Provençal for the bit of waste ground on which the author studies his insects in the natural state.—*Translator's note*.

land. Because wine paid handsomely, they pulled up the forest to plant the vine. Then came the Phylloxera, the vine-stocks perished and the once green table-land is now no more than a desolate stretch where a few tufts of hardy grasses sprout among the pebbles. This waste-land is the Lycosa's paradise: in an hour's time, if need were, I should discover a hundred burrows within a limited range.

These dwellings are pits about a foot deep, perpendicular at first and then bent elbow-wise. The average diameter is an inch. On the edge of the hole stands a kerb, formed of straw, bits and scraps of all sorts and even small pebbles, the size of a hazel-nut. The whole is kept in place and cemented with silk. Often, the Spider confines herself to drawing together the dry blades of the nearest grass, which she ties down with the straps from her spinnerets, without removing the blades from the stems; often, also, she rejects this scaffolding in favour of a masonry constructed of small stones. The nature of the kerb is decided by the nature of the materials within the Lycosa's reach, in the close neighbourhood of the building-yard. There is no selection: everything meets with approval, provided that it be near at hand.

Economy of time, therefore, causes the defensive wall to vary greatly as regards its constituent elements. The height varies also. One enclosure is a turret an inch high; another amounts to a mere rim. All have their parts bound firmly together with silk; and all have the same width as the subterranean channel, of which they are the extension. There is here no difference in diameter between the underground manor and its outwork, nor do we behold, at the opening, the platform which the turret leaves to give free play to the Italian Tarantula's legs. The Black-bellied Tarantula's work takes the form of a well surmounted by its kerb.

When the soil is earthy and homogeneous, the architectural type is free from obstructions and the Spider's dwelling is a cylindrical tube; but, when the site is pebbly, the shape is modified according to the exigencies of the digging. In the second case, the lair is often a rough, winding cave, at intervals along whose inner wall stick blocks of stone avoided in the process of excavation. Whether regular or irregular, the house is plastered to a certain depth with a coat of silk, which prevents earth-slips and facilitates scaling when a prompt exit is required.

Baglivi, in his unsophisticated Latin, teaches us how to catch the Tarantula. I became his *rusticus insidiator*; I waved a spikelet at the entrance of the burrow to imitate the humming of a Bee and attract the attention of the Lycosa, who rushes out, thinking that she is capturing a prey. This method did not succeed with me. The Spider, it is true, leaves her remote apartments and comes a little way up the vertical tube to enquire into the sounds at her door; but the wily animal soon scents a trap; it remains motionless at mid-height and, at the least alarm, goes down again to the branch gallery, where it is invisible.

Léon Dufour's appears to me a better method if it were only practicable in the conditions wherein I find myself. To drive a knife quickly into the ground, across the burrow, so as to cut off the Tarantula's retreat when she is attracted by the spikelet and standing on the upper floor, would be a manoeuvre certain of success, if the soil were favourable. Unfortunately, this is not so in my case: you might as well try to dig a knife into a block of tufa.

Other stratagems become necessary. Here are two which were successful: I recommend them to future Tarantula-hunters. I insert into the burrow, as far down as I can, a stalk with a fleshy spikelet, which the Spider can bite into. I move and turn and twist my bait. The Tarantula, when touched by the intruding body, contemplates self-defence and bites the spikelet. A slight resistance informs my fingers that the animal has fallen into the trap and seized the tip of the stalk in its fangs. I draw it to me, slowly, carefully; the Spider hauls from below, planting her legs against the wall. It comes, it rises. I hide as best I may, when the Spider enters the perpendicular tunnel: if she saw me, she would let go the bait and slip down again. I thus bring her, by degrees, to the orifice. This is the difficult moment. If I continue the gentle movement, the Spider, feeling herself dragged out of her home, would at once run back indoors. It is impossible to get the suspicious animal out by this means. Therefore, when it appears at the level of the ground, I give a sudden pull. Surprised by this foul play, the Tarantula has no time to release her hold; gripping the spikelet, she is thrown some inches away from the burrow. Her capture now becomes an easy matter. Outside her own house, the Lycosa is timid, as though scared, and

GRIPPING THE SPIKELET, SHE IS THROWN SOME INCHES AWAY FROM THE BURROW

hardly capable of running away. To push her with a straw into a paper bag is the affair of a second.

It requires some patience to bring the Tarantula who has bitten into the insidious spikelet to the entrance of the burrow. The following method is quicker: I procure a supply of live Bumble-bees. I put one into a little bottle with a mouth just wide enough to cover the opening of the burrow; and I turn the apparatus thus baited over the said opening. The powerful Bee at first flutters and hums about her glass prison; then, perceiving a burrow similar to that of her family, she enters it without much hesitation. She is extremely ill-advised: while she goes down, the Spider comes up; and the meeting takes place in the perpendicular passage. For a few moments, the ear perceives a sort of death-song: it is the humming of the Bumble-bee, protesting against the reception given her. This is followed by a long silence. Then I remove the bottle and dip a long-jawed forceps into the pit. I withdraw the Bumble-bee, motionless, dead, with hanging proboscis. A terrible tragedy must have happened. The Spider follows, refusing to let go so rich a booty. Game and huntress are brought to the orifice. Sometimes, mistrustful, the Lycosa goes in again; but we have only to leave the Bumble-bee on the threshold of the door, or even a few inches away, to see her reappear, issue from her fortress and daringly recapture her prey. This is the moment: the house is closed with the finger, or a pebble and, as Baglivi says, '*captatur tamen ista a rustico insidiatore*,' to which I will add, '*adjuvante Bombo.*' [1]

The object of these hunting methods was not exactly to obtain Tarantulae; I had not the least wish to rear the Spider in a bottle. I was interested in a different matter. Here, thought I, is an ardent huntress, living solely by her trade. She does not prepare preserved foodstuffs for her offspring;[2] she herself feeds on the prey which she catches. She is not a 'paralyzer,' [3] who cleverly spares her quarry so as to leave it a glimmer of life and keep it fresh for weeks at a time; she is a killer, who makes a meal off her capture on the spot. With her, there is no methodical vivisection, which destroys movement without entirely destroying life, but absolute death, as sudden as

---

[1] 'Thanks to the Bumble-bee.'
[2] Like the Dung-beetles.—Translator's Note.
[3] Like the Solitary Wasps.—Translator's Note.

possible, which protects the assailant from the counter-attacks of the assailed.

Her game, moreover, is essentially bulky and not always of the most peaceful character. This Diana, ambushed in her tower, needs a prey worthy of her prowess. The big Grasshopper, with the powerful jaws; the irascible Wasp; the Bee, the Bumble-bee and other wearers of poisoned daggers must fall into the ambuscade from time to time. The duel is nearly equal in point of weapons. To the venomous fangs of the Lycosa the Wasp opposes her venomous stiletto. Which of the two bandits shall have the best of it? The struggle is a hand-to-hand one. The Tarantula has no secondary means of defence, no cord to bind her victim, no trap to subdue her. When the Epeira, or Garden Spider, sees an insect entangled in her great upright web, she hastens up and covers the captive with corded meshes and silk ribbons by the armful, making all resistance impossible. When the prey is solidly bound, a prick is carefully administered with the poison-fangs; then the Spider retires, waiting for the death-throes to calm down, after which the huntress comes back to the game. In these conditions, there is no serious danger.

In the case of the Lycosa, the job is riskier. She has naught to serve her but her courage and her fangs and is obliged to leap upon the formidable prey, to master it by her dexterity, to annihilate it, in a measure, by her swift-slaying talent.

Annihilate is the word: the Bumble-bees whom I draw from the fatal hole are a sufficient proof. As soon as that shrill buzzing, which I called the death-song, ceases, in vain I hasten to insert my forceps: I always bring out the insect dead, with slack proboscis and limp legs. Scarce a few quivers of those legs tell me that it is a quite recent corpse. The Bumble-bee's death is instantaneous. Each time that I take a fresh victim from the terrible slaughter-house, my surprise is renewed at the sight of its sudden immobility.

Nevertheless, both animals have very nearly the same strength; for I choose my Bumble-bees from among the largest (*Bombus hortorum* and *B. terrestris*). Their weapons are almost equal: the Bee's dart can bear comparison with the Spider's fangs; the sting of the first seems to me as formidable as the bite of the second. How comes it that the Tarantula always has the upper hand and this moreover in a very short conflict, whence she emerges unscathed?

There must certainly be some cunning strategy on her part. Subtle though her poison may be, I cannot believe that its mere injection, at any point whatever of the victim, is enough to produce so prompt a catastrophe. The ill-famed rattlesnake does not kill so quickly, takes hours to achieve that for which the Tarantula does not require a second. We must, therefore, look for an explanation of this sudden death to the vital importance of the point attacked by the Spider, rather than to the virulence of the poison.

What is this point? It is impossible to recognize it on the Bumble-bees. They enter the burrow; and the murder is committed far from sight. Nor does the lens discover any wound upon the corpse, so delicate are the weapons that produce it. One would have to see the two adversaries engage in a direct contest. I have often tried to place a Tarantula and a Bumble-bee face to face in the same bottle. The two animals mutually flee each other, each being as much upset as the other at its captivity. I have kept them together for twenty-four hours, without aggressive display on either side. Thinking more of their prison than of attacking each other, they temporize, as though indifferent. The experiment has always been fruitless. I have succeeded with Bees and Wasps, but the murder has been committed at night and has taught me nothing. I would find both insects, next morning, reduced to a jelly under the Spider's mandibles. A weak prey is a mouthful which the Spider reserves for the calm of the night. A prey capable of resistance is not attacked in captivity. The prisoner's anxiety cools the hunter's ardour.

The arena of a large bottle enables each athlete to keep out of the other's way, respected by her adversary, who is respected in her turn. Let us reduce the lists, diminish the enclosure. I put Bumble-bee and Tarantula into a test-tube that has only room for one at the bottom. A lively brawl ensues, without serious results. If the Bumble-bee be underneath, she lies down on her back and with her legs wards off the other as much as she can. I do not see her draw her sting. The Spider, meanwhile, embracing the whole circumference of the enclosure with her long legs, hoists herself a little upon the slippery surface and removes herself as far as possible from her adversary. There, motionless, she awaits events, which are soon disturbed by the fussy Bumble-bee. Should the latter occupy the upper position, the Tarantula protects herself by drawing up her legs, which keep the enemy at a distance. In short, save for sharp scuffles when the two champions are in touch, nothing happens that deserves attention. There is no duel to the death in the narrow arena of the test-tube, any more than in the wider lists afforded by the bottle. Utterly timid once she is away from home, the Spider obstinately refuses the battle; nor will the Bumble-bee, giddy though she be, think of striking the first blow. I abandon experiments in my study.

We must go direct to the spot and force the duel upon the Tarantula, who is full of pluck in her own stronghold. Only, instead of the Bumble-bee, who enters the burrow and conceals her death from our eyes, it is necessary to substitute another adversary, less inclined to penetrate underground. There abounds in the garden, at this moment, on the flowers of the common clary, one of the largest and most powerful Bees that haunt my district, the Carpenter-bee (*Xylocopa violacea*), clad in black velvet, with wings of purple gauze. Her size, which is nearly an inch, exceeds that of the Bumble-bee. Her sting is excruciating and produces a swelling that long continues painful. I have very exact memories on this subject, memories that have cost me dear. Here indeed is an antagonist worthy of the Tarantula, if I succeed in inducing the Spider to accept her. I place a certain number, one by one, in bottles small in capacity, but having a wide neck capable of surrounding the entrance to the burrow.

As the prey which I am about to offer is capable of overawing the huntress, I select from among the Tarantulae the lustiest, the boldest, those most stimulated by hunger. The spikeleted stalk is pushed into the burrow. When the Spider hastens up at once, when she is of a good size, when she climbs boldly to the aperture of her dwelling, she is admitted to the tourney; otherwise, she is refused. The bottle, baited with a Carpenter-bee, is placed upside down over the door of one of the elect. The Bee buzzes gravely in her glass bell; the huntress mounts from the recesses of the cave; she is on the threshold, but inside; she looks; she waits. I also wait. The quarters, the half-hours pass: nothing. The Spider goes down again: she has probably judged the attempt too dangerous. I move to a second, a third, a fourth burrow: still nothing; the huntress refuses to leave her lair.

Fortune at last smiles upon my patience, which has been heavily tried by all these prudent retreats and particularly by the fierce heat of the dog-days. A Spider suddenly rushes from her hole: she has been rendered warlike, doubtless, by prolonged abstinence. The tragedy that happens under the cover of the bottle lasts for but the twinkling of an eye. It is over: the sturdy Carpenter-bee is dead. Where did the murderess strike her? That is easily ascertained: the Tarantula has not let go; and her fangs are planted in the nape of the neck. The assassin has the knowledge which I suspected: she has made for the essentially vital centre, she has stung the insect's cervical ganglia with her poison-fangs. In short, she has bitten the only point a lesion in which produces sudden death. I was delighted with this murderous skill, which made amends for the blistering which my skin received in the sun.

Once is not custom: one swallow does not make a summer. Is what I have just seen due to accident or to premeditation? I turn to other Lycosae. Many, a deal too many for my patience, stubbornly refuse to dart from their haunts in order to attack the Carpenter-bee. The formidable quarry is too much for their daring. Shall not hunger, which brings the wolf from the wood, also bring the Tarantula out of her hole? Two, apparently more famished than the rest, do at last pounce upon the Bee and repeat the scene of murder before my eyes. The prey, again bitten in the neck, exclusively in the neck, dies on the instant. Three murders, perpetrated in my presence under identical conditions, represent the fruits of my experiment pursued, on two occasions, from eight o'clock in the morning until twelve midday.

I had seen enough. The quick insect-killer had taught me her trade as had the paralyzer[1] before her: she had shown me that she is thoroughly versed in the art of the butcher of the Pampas[2]. The Tarantula is an accomplished *desnucador*. It remained to me to confirm the open-air experiment with experiments in the privacy of my study. I therefore got together a menagerie of these poisonous

---

[1]Such as the Hairy Ammophila, the Cerceris and the Languedocian Sphex, Digger-wasps described in other of the author's essays.—*Translator's Note*.

[2]The *desnucador*, the Argentine slaughterman whose methods of slaying cattle are detailed in the author's essay entitled, The Theory of Instinct.—*Translator's Note*.

Spiders, so as to judge of the virulence of their venom and its effect according to the part of the body injured by the fangs. A dozen bottles and test-tubes received the prisoners, whom I captured by the methods known to the reader. To one inclined to scream at the sight of a Spider, my study, filled with odious Lycosae, would have presented a very uncanny appearance.

Though the Tarantula scorns or rather fears to attack an adversary placed in her presence in a bottle, she scarcely hesitates to bite what is thrust beneath her fangs. I take her by the thorax with my forceps and present to her mouth the animal which I wish stung. Forthwith, if the Spider be not already tired by experiments, the fangs are raised and inserted. I first tried the effects of the bite upon the Carpenter-bee. When struck in the neck, the Bee succumbs at once. It was the lightning death which I witnessed on the threshold of the burrows. When struck in the abdomen and then placed in a large bottle that leaves its movements free, the insect seems, at first, to have suffered no serious injury. It flutters about and buzzes. But half an hour has not elapsed before death is imminent. The insect lies motionless upon its back or side. At most, a few movements of the legs, a slight pulsation of the belly, continuing till the morrow, proclaim that life has not yet entirely departed. Then everything ceases: the Carpenter-bee is a corpse.

The importance of this experiment compels our attention. When stung in the neck, the powerful Bee dies on the spot; and the Spider has not to fear the dangers of a desperate struggle. Stung elsewhere, in the abdomen, the insect is capable, for nearly half an hour, of making use of its dart, its mandibles, its legs; and woe to the Lycosa whom the stiletto reaches. I have seen some who, stabbed in the mouth while biting close to the sting, died of the wound within the twenty-four hours. That dangerous prey, therefore, requires instantaneous death, produced by the injury to the nerve-centres of the neck; otherwise, the hunter's life would often be in jeopardy.

The Grasshopper order supplied me with a second series of victims: Green Grasshoppers as long as one's finger, large-headed Locusts, Ephippigerae. The same result follows when these are bitten in the neck: lightning death. When injured elsewhere, notably in the abdomen, the subject of the experiment resists for some time. I have seen a Grasshopper, bitten in the belly, cling firmly for fifteen hours to the smooth, upright wall of the glass bell that constituted his prison. At last, he dropped off and died. Where the Bee, that delicate organism, succumbs in less than half an hour, the Grasshopper, coarse ruminant that he is, resists for a whole day. Put aside these differences, caused by unequal degrees of organic sensitiveness, and we sum up as follows: when bitten by the Tarantula in the neck, an insect, chosen from among the largest, dies on the spot; when bitten elsewhere, it perishes also, but after a lapse of time which varies considerably in the different entomological orders.

This explains the long hesitation of the Tarantula, so wearisome to the experimenter when he presents to her, at the entrance to the burrow, a rich, but dangerous prey. The majority refuse to fling themselves upon the Carpenter-bee. The fact is that a quarry of this kind cannot be seized recklessly: the huntress who missed her stroke by biting at random would do so at the risk of her life. The nape of the neck alone possesses the desired vulnerability. The adversary must be nipped there and no elsewhere. Not to floor her at once would mean to irritate her and make her more dangerous than ever. The Spider is well aware of this. In the safe shelter of her threshold, therefore, prepared to beat a quick retreat if necessary, she watches for the favourable moment; she waits for the big Bee to face her, when the neck is easily grabbed. If this condition of success offer, she leaps out and acts; if not, weary of the violent evolutions of the quarry, she retires indoors. And that, no doubt, is why it took me two sittings of four hours apiece to witness three assassinations.

Formerly, instructed by the paralysing Wasps, I had myself tried to produce paralysis by injecting a drop of ammonia into the thorax of those insects, such as Weevils, Buprestes,[1] and Dung-beetles, whose compact nervous system assists this physiological operation. I showed myself a ready pupil to my masters' teaching and used to paralyze a Buprestis or a Weevil almost as well as a Cerceris[2] could have done. Why should I not to-day imitate that expert butcher, the Tarantula? With the point of a fine needle, I inject a tiny drop of ammonia at the base of the skull of a Carpenter-bee or a Grasshopper. The insect succumbs then and there, without any

[1] A genus of Beetles.—*Translator's Note*.
[2] A species of Digger-wasp.—Translator's Note.

other movement than wild convulsions. When attacked by the acrid fluid, the cervical ganglia cease to do their work; and death ensues. Nevertheless, this death is not immediate; the throes last for some time. The experiment is not wholly satisfactory as regards suddenness. Why? Because the liquid which I employ, ammonia, cannot be compared, for deadly efficacy, with the Lycosa's poison, a pretty formidable poison, as we shall see.

I make a Tarantula bite the leg of a young, well-fledged Sparrow, ready to leave the nest. A drop of blood flows; the wounded spot is surrounded by a reddish circle, changing to purple. The bird almost immediately loses the use of its leg, which drags, with the toes doubled in; it hops upon the other. Apart from this, the patient does not seem to trouble much about his hurt; his appetite is good. My daughters feed him on Flies, bread-crumb, apricot-pulp. He is sure to get well, he will recover his strength; the poor victim of the curiosity of science will be restored to liberty. This is the wish, the intention of us all. Twelve hours later, the hope of a cure increases; the invalid takes nourishment readily; he clamours for it, if we keep him waiting. But the leg still drags. I set this down to a temporary paralysis which will soon disappear. Two days after, he refuses his food. Wrapping himself in his stoicism and his rumpled feathers, the Sparrow hunches into a ball, now motionless, now twitching. My girls take him in the hollow of their hands and warm him with their breath. The spasms become more frequent. A gasp proclaims that all is over. The bird is dead.

There was a certain coolness among us at the evening-meal. I read mute reproaches, because of my experiment, in the eyes of my home-circle; I read an unspoken accusation of cruelty all around me. The death of the unfortunate Sparrow had saddened the whole family. I myself was not without some remorse of conscience: the poor result achieved seemed to me too dearly bought. I am not made of the stuff of those who, without turning a hair, rip up live Dogs to find out nothing in particular.

Nevertheless, I had the courage to start afresh, this time on a Mole caught ravaging a bed of lettuces. There was a danger lest my captive, with his famished stomach, should leave things in doubt, if we had to keep him for a few days. He might die not of his wound, but of inanition, if I did not succeed in giving him suitable food, fairly plentiful and dispensed at fairly frequent intervals. In that case, I ran a risk of ascribing to the poison what might well be the result of starvation. I must therefore begin by finding out if it was possible for me to keep the Mole alive in captivity. The animal was put into a large receptacle from which it could not get out and fed on a varied diet of insects—Beetles, Grasshoppers, especially Cicadae[1] —which it crunched up with an excellent appetite. Twenty-four hours of this regimen convinced me that the Mole was making the best of the bill of fare and taking kindly to his captivity.

I make the Tarantula bite him at the tip of the snout. When replaced in his cage, the Mole keeps on scratching his nose with his broad paws. The thing seems to burn, to itch. Henceforth, less and less of the provision of Cicadae is consumed; on the evening of the following day, it is refused altogether. About thirty-six hours after being bitten, the Mole dies during the night and certainly not from inanition, for there are still half a dozen live Cicadae in the receptacle, as well as a few Beetles.

The bite of the Black-bellied Tarantula is therefore dangerous to other animals than insects: it is fatal to the Sparrow, it is fatal to the Mole. Up to what point are we to generalize? I do not know, because my enquiries extended no further. Nevertheless, judging from the little that I saw, it appears to me that the bite of this Spider is not an accident which man can afford to treat lightly. This is all that I have to say to the doctors.

To the philosophical entomologists I have something else to say: I have to call their attention to the consummate knowledge of the insect-killers, which vies with that of the paralyzers. I speak of insect-killers in the plural, for the Tarantula must share her deadly art with a host of other Spiders, especially with those who hunt without nets. These insect-killers, who live on their prey, strike the game dead instantaneously by stinging the nerve-centres of the neck; the paralyzers, on the other hand, who wish to keep the food fresh for their larvae, destroy the power of movement by stinging the game in the other nerve-centres. Both of them attack the nervous chain, but they select the point according to the object to be attained. If death be desired, sudden death, free from danger to

---

[1] The Cicada is the Cigale, an insect akin to the Grasshopper and found more particularly in the South of France.—*Translator's Note*.

the huntress, the insect is attacked in the neck; if mere paralysis be required, the neck is respected and the lower segments—sometimes one alone, sometimes three, sometimes all or nearly all, according to the special organization of the victim—receive the dagger-thrust.

Even the paralyzers, at least some of them, are acquainted with the immense vital importance of the nerve-centres of the neck. We have seen the Hairy Ammophila munching the caterpillar's brain, the Languedocian Sphex munching the brain of the Ephippigera, with the object of inducing a passing torpor. But they simply squeeze the brain and do even this with a wise discretion; they are careful not to drive their sting into this fundamental centre of life; not one of them ever thinks of doing so, for the result would be a corpse which the larva would despise. The Spider, on the other hand, inserts her double dirk there and there alone; any elsewhere it would inflict a wound likely to increase resistance through irritation. She wants a venison for consumption without delay and brutally thrusts her fangs into the spot which the others so conscientiously respect.

If the instinct of these scientific murderers is not, in both cases, an inborn predisposition, inseparable from the animal, but an acquired habit, then I rack my brain in vain to understand how that habit can have been acquired. Shroud these facts in theoretic mists as much as you will, you shall never succeed in veiling the glaring evidence which they afford of a pre-established order of things.

CHAPTER II

THE BANDED EPEIRA

IN the inclement season of the year, when the insect has nothing to do and retires to winter quarters, the observer profits by the mildness of the sunny nooks and grubs in the sand, lifts the stones, searches the brushwood; and often he is stirred with a pleasurable excitement, when he lights upon some ingenious work of art, discovered unawares. Happy are the simple of heart whose ambition is satisfied with such treasure-trove! I wish them all the joys which it has brought me and which it will continue to bring me, despite the vexations of life, which grow ever more bitter as the years follow their swift downward course.

Should the seekers rummage among the wild grasses in the osier-beds and copses, I wish them the delight of finding the wonderful object that, at this moment, lies before my eyes. It is the work of a Spider, the nest of the Banded Epeira (*Epeira fasciata*, LATR.).

A Spider is not an insect, according to the rules of classification; and as such the Epeira seems out of place here.[1] A fig for systems! It is immaterial to the student of instinct whether the animal have eight legs instead of six, or pulmonary sacs instead of air-tubes. Besides, the Araneida belong to the group of segmented animals, organized in sections placed end to end, a structure to which the terms 'insect' and 'entomology' both refer.

Formerly, to describe this group, people said 'articulate animals,' an expression which possessed the drawback of not jarring on the ear and of being understood by all. This is out of date. Nowadays, they use the euphonious term 'Arthropoda.' And to think that there are men who question the existence of progress! Infidels! Say,

---

[1] The generic title of the work from which these essays are taken is Entomological Memories, or, Studies relating to the Instinct and Habits of Insects.—*Translator's Note*.

'articulate,' first; then roll out, 'Arthropoda;' and you shall see whether zoological science is not progressing!

In bearing and colouring, *Epeira fasciata* is the handsomest of the Spiders of the South. On her fat belly, a mighty silk-warehouse nearly as large as a hazel-nut, are alternate yellow, black and silver sashes, to which she owes her epithet of Banded. Around that portly abdomen, the eight long legs, with their dark- and pale-brown rings, radiate like spokes.

Any small prey suits her; and, as long as she can find supports for her web, she settles wherever the Locust hops, wherever the Fly hovers, wherever the Dragon-fly dances or the Butterfly flits. As a rule, because of the greater abundance of game, she spreads her toils across some brooklet, from bank to bank among the rushes. She also stretches them, but not assiduously, in the thickets of evergreen oak, on the slopes with the scrubby greenswards, dear to the Grasshoppers.

Her hunting-weapon is a large upright web, whose outer boundary, which varies according to the disposition of the ground, is fastened to the neighbouring branches by a number of moorings. The structure is that adopted by the other weaving Spiders. Straight threads radiate at equal intervals from a central point. Over this framework runs a continuous spiral thread, forming chords, or cross-bars, from the centre to the circumference. It is magnificently large and magnificently symmetrical.

In the lower part of the web, starting from the centre, a wide opaque ribbon descends zigzag-wise across the radii. This is the Epeira's trade-mark, the flourish of an artist initialling his creation. '*Fecit*So-and-so,' she seems to say, when giving the last throw of the shuttle to her handiwork.

That the Spider feels satisfied when, after passing and repassing from spoke to spoke, she finishes her spiral, is beyond a doubt: the work achieved ensures her food for a few days to come. But, in this particular case, the vanity of the spinstress has naught to say to the matter: the strong silk zigzag is added to impart greater firmness to the web.

Increased resistance is not superfluous, for the net is sometimes exposed to severe tests. The Epeira cannot pick and choose her prizes. Seated motionless in the centre of her web, her eight legs wide-spread to feel the shaking of the network in any direction, she waits for what luck will bring her: now some giddy weakling unable to control its flight, anon some powerful prey rushing headlong with a reckless bound.

The Locust in particular, the fiery Locust, who releases the spring of his long shanks at random, often falls into the trap. One imagines that his strength ought to frighten the Spider; the kick of his spurred levers should enable him to make a hole, then and there, in the web and to get away. But not at all. If he does not free himself at the first effort, the Locust is lost.

Turning her back on the game, the Epeira works all her spinnerets, pierced like the rose of a watering-pot, at one and the same time. The silky spray is gathered by the hind-legs, which are longer than the others and open into a wide arc to allow the stream to spread. Thanks to this artifice, the Epeira this time obtains not a thread, but an iridescent sheet, a sort of clouded fan wherein the component threads are kept almost separate. The two hind-legs fling this shroud gradually, by rapid alternate armfuls, while, at the same time, they turn the prey over and over, swathing it completely.

The ancient *retiarius*, when pitted against a powerful wild beast, appeared in the arena with a rope-net folded over his left shoulder. The animal made its spring. The man, with a sudden movement of his right arm, cast the net after the manner of the fishermen; he covered the beast and tangled it in the meshes. A thrust of the trident gave the quietus to the vanquished foe.

The Epeira acts in like fashion, with this advantage, that she is able to renew her armful of fetters. Should the first not suffice, a second instantly follows and another and yet another, until the reserves of silk become exhausted.

When all movement ceases under the snowy winding-sheet, the Spider goes up to her bound prisoner. She has a better weapon than the *bestiarius'* trident: she has her poison-fangs. She gnaws at the Locust, without undue persistence, and then withdraws, leaving the torpid patient to pine away.

Soon she comes back to her motionless head of game: she sucks it, drains it, repeatedly changing her point of attack. At last, the clean-bled remains are flung out of the net and the Spider returns to her ambush in the centre of the web.

What the Epeira sucks is not a corpse, but a numbed body. If I remove the Locust immediately after he has been bitten and release him from the silken sheath, the patient recovers his strength to such an extent that he seems, at first, to have suffered no injury. The Spider, therefore, does not kill her capture before sucking its juices; she is content to deprive it of the power of motion by producing a state of torpor. Perhaps this kindlier bite gives her greater facility in working her pump. The humours, if stagnant, in a corpse, would not respond so readily to the action of the sucker; they are more easily extracted from a live body, in which they move about.

The Epeira, therefore, being a drinker of blood, moderates the virulence of her sting, even with victims of appalling size, so sure is she of her retiarian art. The long-legged Tryxalis[1], the corpulent Grey Locust, the largest of our Grasshoppers are accepted without hesitation and sucked dry as soon as numbed. Those giants, capable of making a hole in the net and passing through it in their impetuous onrush, can be but rarely caught. I myself place them on the web. The Spider does the rest. Lavishing her silky spray, she swathes them and then sucks the body at her ease. With an increased expenditure of the spinnerets, the very biggest game is mastered as successfully as the everyday prey.

I have seen even better than that. This time, my subject is the Silky Epeira (*Epeira sericea*, OLIV.), with a broad, festooned, silvery abdomen. Like that of the other, her web is large, upright and 'signed' with a zigzag ribbon. I place upon it a Praying Mantis[2], a well-developed specimen, quite capable of changing rôles, should circumstances permit, and herself making a meal off her assailant. It is a question no longer of capturing a peaceful Locust, but a fierce and powerful ogre, who would rip open the Epeira's paunch with one blow of her harpoons.

Will the Spider dare? Not immediately. Motionless in the centre of her net, she consults her strength before attacking the formidable quarry; she waits until the struggling prey has its claws more thickly

[1] A species of Grasshopper.—*Translator's Note.*

[2] An insect akin to the Locusts and Crickets, which, when at rest, adopts an attitude resembling that of prayer. When attacking, it assumes what is known as 'the spectral attitude.' Its forelegs form a sort of saw-like or barbed harpoons. Cf. Social Life in the Insect World, by J. H. Fabre, translated by Bernard Miall: chaps. v. to vii.—*Translator's Note.*

entangled. At last, she approaches. The Mantis curls her belly; lifts her wings like vertical sails; opens her saw-toothed arm-pieces; in short, adopts the spectral attitude which she employs when delivering battle.

The Spider disregards these menaces. Spreading wide her spinnerets, she pumps out sheets of silk which the hind-legs draw out, expand and fling without stint in alternate armfuls. Under this shower of threads, the Mantis' terrible saws, the lethal legs, quickly disappear from sight, as do the wings, still erected in the spectral posture.

Meanwhile, the swathed one gives sudden jerks, which make the Spider fall out of her web. The accident is provided for. A safety-cord, emitted at the same instant by the spinnerets, keeps the Epeira hanging, swinging in space. When calm is restored, she packs her cord and climbs up again. The heavy paunch and the hind-legs are now bound. The flow slackens, the silk comes only in thin sheets. Fortunately, the business is done. The prey is invisible under the thick shroud.

The Spider retires without giving a bite. To master the terrible quarry, she has spent the whole reserves of her spinning-mill, enough to weave many good-sized webs. With this heap of shackles, further precautions are superfluous.

After a short rest in the centre of the net, she comes down to dinner. Slight incisions are made in different parts of the prize, now here, now there; and the Spider puts her mouth to each and sucks the blood of her prey. The meal is long protracted, so rich is the dish. For ten hours, I watch the insatiable glutton, who changes her point of attack as each wound sucked dries up. Night comes and robs me of the finish of the unbridled debauch. Next morning, the drained Mantis lies upon the ground. The Ants are eagerly devouring the remains.

The eminent talents of the Epeirae are displayed to even better purpose in the industrial business of motherhood than in the art of the chase. The silk bag, the nest, in which the Banded Epeira houses her eggs, is a much greater marvel than the bird's nest. In shape, it is an inverted balloon, nearly the size of a Pigeon's egg. The top tapers like a pear and is cut short and crowned with a scalloped rim, the corners of which are lengthened by means of moorings that fasten

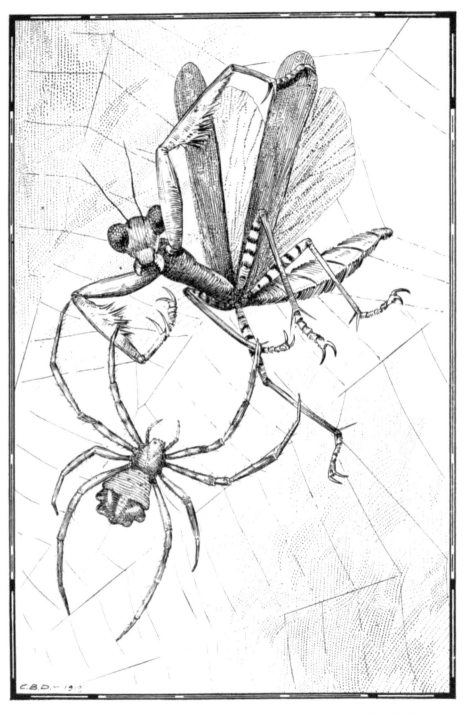

THE MANTIS... ADOPTS THE SPECTRAL ATTITUDE
WHICH SHE EMPLOYS WHEN DELIVERING BATTLE.

the object to the adjoining twigs. The whole, a graceful ovoid, hangs straight down, amid a few threads that steady it.

The top is hollowed into a crater closed with a silky padding. Every other part is contained in the general wrapper, formed of thick, compact white satin, difficult to break and impervious to moisture. Brown and even black silk, laid out in abroad ribbons, in spindle-shaped patterns, in fanciful meridian waves, adorns the upper portion of the exterior. The part played by this fabric is self-evident: it is a waterproof cover which neither dew nor rain can penetrate.

Exposed to all the inclemencies of the weather, among the dead grasses, close to the ground, the Epeira's nest has also to protect its contents from the winter cold. Let us cut the wrapper with our scissors. Underneath, we find a thick layer of reddish-brown silk, not worked into a fabric this time, but puffed into an extra-fine wadding. It is a fleecy cloud, an incomparable quilt, softer than any swan's-down. This is the screen set up against loss of heat.

And what does this cosy mass protect? See: in the middle of the eiderdown hangs a cylindrical pocket, round at the bottom, cut square at the top and closed with a padded lid. It is made of extremely fine satin; it contains the Epeira's eggs, pretty little orange-coloured beads, which, glued together, form a globule the size of a pea. This is the treasure to be defended against the asperities of the winter.

Now that we know the structure of the work, let us try to see in what manner the spinstress sets about it. The observation is not an easy one, for the Banded Epeira is a night-worker. She needs nocturnal quiet in order not to go astray amid the complicated rules that guide her industry. Now and again, at very early hours in the morning, I have happened to catch her working, which enables me to sum up the progress of the operations.

My subjects are busy in their bell-shaped cages, at about the middle of August. A scaffolding is first run up, at the top of the dome; it consists of a few stretched threads. The wire trellis represents the twigs and the blades of grass which the Spider, if at liberty, would have used as suspension-points. The loom works on this shaky support. The Epeira does not see what she is doing; she turns her back on her task. The machinery is so well put together that the whole thing goes automatically.

The tip of the abdomen sways, a little to the right, a little to the left, rises and falls, while the Spider moves slowly round and round. The thread paid out is single. The hind-legs draw it out and place it in position on that which is already done. Thus is formed a satin receptacle the rim of which is gradually raised until it becomes a bag about a centimetre deep[1]. The texture is of the daintiest. Guy-ropes bind it to the nearest threads and keep it stretched, especially at the mouth.

Then the spinnerets take a rest and the turn of the ovaries comes. A continuous shower of eggs falls into the bag, which is filled to the top. The capacity of the receptacle has been so nicely calculated that there is room for all the eggs, without leaving any space unoccupied. When the Spider has finished and retires, I catch a momentary glimpse of the heap of orange-coloured eggs; but the work of the spinnerets is at once resumed.

The next business is to close the bag. The machinery works a little differently. The tip of the belly no longer sways from side to side. It sinks and touches a point; it retreats, sinks again and touches another point, first here, then there, describing inextricable zigzags. At the same time, the hind-legs tread the material emitted. The result is no longer a stuff, but a felt, a blanketing.

Around the satin capsule, which contains the eggs, is the eiderdown destined to keep out the cold. The youngsters will bide for some time in this soft shelter, to strengthen their joints and prepare for the final exodus. It does not take long to make. The spinning-mill suddenly alters the raw material: it was turning out white silk; it now furnishes reddish-brown silk, finer than the other and issuing in clouds which the hind-legs, those dexterous carders, beat into a sort of froth. The egg-pocket disappears, drowned in this exquisite wadding.

The balloon-shape is already outlined; the top of the work tapers to a neck. The Spider, moving up and down, tacking first to one side and then to the other, from the very first spray marks out the graceful form as accurately as though she carried a compass in her abdomen.

[1] .39 inch.—*Translator's Note.*

Then, once again, with the same suddenness, the material changes. The white silk reappears, wrought into thread. This is the moment to weave the outer wrapper. Because of the thickness of the stuff and the density of its texture, this operation is the longest of the series.

First, a few threads are flung out, hither and thither, to keep the layer of wadding in position. The Epeira takes special pains with the edge of the neck, where she fashions an indented border, the angles of which, prolonged with cords or lines, form the main support of the building. The spinnerets never touch this part without giving it, each time, until the end of the work, a certain added solidity, necessary to secure the stability of the balloon. The suspensory indentations soon outline a crater which needs plugging. The Spider closes the bag with a padded stopper similar to that with which she sealed the egg-pocket.

When these arrangements are made, the real manufacture of the wrapper begins. The Spider goes backwards and forwards, turns and turns again. The spinnerets do not touch the fabric. With a rhythmical, alternate movement, the hind-legs, the sole implements employed, draw the thread, seize it in their combs and apply it to the work, while the tip of the abdomen sways methodically to and fro.

In this way, the silken fibre is distributed in an even zigzag, of almost geometrical precision and comparable with that of the cotton thread which the machines in our factories roll so neatly into balls. And this is repeated all over the surface of the work, for the Spider shifts her position a little at every moment.

At fairly frequent intervals, the tip of the abdomen is lifted to the mouth of the balloon; and then the spinnerets really touch the fringed edge. The length of contact is even considerable. We find, therefore, that the thread is stuck in this star-shaped fringe, the foundation of the building and the crux of the whole, while every elsewhere it is simply laid on, in a manner determined by the movements of the hind-legs. If we wished to unwind the work, the thread would break at the margin; at any other point, it would unroll.

The Epeira ends her web with a dead-white, angular flourish; she ends her nest with brown mouldings, which run down, irregularly, from the marginal junction to the bulging middle. For this purpose, she makes use, for the third time, of a different silk; she now produces silk of a dark hue, varying from russet to black. The spinnerets distribute the material with a wide longitudinal swing, from pole to pole; and the hind-legs apply it in capricious ribbons. When this is done, the work is finished. The Spider moves away with slow strides, without giving a glance at the bag. The rest does not interest her: time and the sun will see to it.

She felt her hour at hand and came down from her web. Near by, in the rank grass, she wove the tabernacle of her offspring and, in so doing, drained her resources. To resume her hunting-post, to return to her web would be useless to her: she has not the wherewithal to bind the prey. Besides, the fine appetite of former days has gone. Withered and languid, she drags out her existence for a few days and, at last, dies. This is how things happen in my cages; this is how they must happen in the brushwood.

The Silky Epeira (*Epeira sericea*, OLIV.) excels the Banded Epeira in the manufacture of big hunting-nets, but she is less gifted in the art of nest-building. She gives her nest the inelegant form of an obtuse cone. The opening of this pocket is very wide and is scalloped into lobes by which the edifice is slung. It is closed with a large lid, half satin, half swan's-down. The rest is a stout white fabric, frequently covered with irregular brown streaks.

The difference between the work of the two Epeirae does not extend beyond the wrapper, which is an obtuse cone in the one case and a balloon in the other. The same internal arrangements prevail behind this frontage: first, a flossy quilt; next, a little keg in which the eggs are packed. Though the two Spiders build the outer wall according to special architectural rules, they both employ the same means as a protection against the cold.

As we see, the egg-bag of the Epeirae, particularly that of the Banded Epeira, is an important and complex work. Various materials enter into its composition: white silk, red silk, brown silk; moreover, these materials are worked into dissimilar products: stout cloth, soft eiderdown, dainty satinette, porous felt. And all of this comes from the same workshop that weaves the hunting-net, warps the zigzag ribbon-band and casts an entangling shroud over the prey.

What a wonderful silk-factory it is! With a very simple and never-varying plant, consisting of the hind-legs and the spinnerets, it produces, by turns, rope-maker's, spinner's, weaver's, ribbon-maker's and fuller's work. How does the Spider direct an establishment of this kind? How does she obtain, at will, skeins of diverse hues and grades? How does she turn them out, first in this fashion, then in that? I see the results, but I do not understand the machinery and still less the process. It beats me altogether.

The Spider also sometimes loses her head in her difficult trade, when some trouble disturbs the peace of her nocturnal labours. I do not provoke this trouble myself, for I am not present at those unseasonable hours. It is simply due to the conditions prevailing in my menagerie.

In their natural state, the Epeirae settle separately, at long distances from one another. Each has her own hunting-grounds, where there is no reason to fear the competition that would result from the close proximity of the nets. In my cages, on the other hand, there is cohabitation. In order to save space, I lodge two or three Epeirae in the same cage. My easy-going captives live together in peace. There is no strife between them, no encroaching on the neighbour's property. Each of them weaves herself a rudimentary web, as far from the rest as possible, and here, rapt in contemplation, as though indifferent to what the others are doing, she awaits the hop of the Locust.

Nevertheless, these close quarters have their drawbacks when laying-time arrives. The cords by which the different establishments are hung interlace and criss-cross in a confused network. When one of them shakes, all the others are more or less affected. This is enough to distract the layer from her business and to make her do silly things. Here are two instances.

A bag has been woven during the night. I find it, when I visit the cage in the morning, hanging from the trellis-work and completed. It is perfect, as regards structure; it is decorated with the regulation black meridian curves. There is nothing missing, nothing except the essential thing, the eggs, for which the spinstress has gone to such expense in the matter of silks. Where are the eggs? They are not in the bag, which I open and find empty. They are lying on the ground below, on the sand in the pan, utterly unprotected.

Disturbed at the moment of discharging them, the mother has missed the mouth of the little bag and dropped them on the floor. Perhaps even, in her excitement, she came down from above and, compelled by the exigencies of the ovaries, laid her eggs on the first support that offered. No matter: if her Spider brain contains the least gleam of sense, she must be aware of the disaster and is therefore bound at once to abandon the elaborate manufacture of a now superfluous nest.

Not at all: the bag is woven around nothing, as accurate in shape, as finished in structure as under normal conditions. The absurd perseverance displayed by certain Bees, whose egg and provisions I used to remove[1], is here repeated without the slightest interference from me. My victims used scrupulously to seal up their empty cells. In the same way, the Epeira puts the eiderdown quilting and the taffeta wrapper round a capsule that contains nothing.

Another, distracted from her work by some startling vibration, leaves her nest at the moment when the layer of red-brown wadding is being completed. She flees to the dome, at a few inches above her unfinished work, and spends upon a shapeless mattress, of no use whatever, all the silk with which she would have woven the outer wrapper if nothing had come to disturb her.

Poor fool! You upholster the wires of your cage with swan's-down and you leave the eggs imperfectly protected. The absence of the work already executed and the hardness of the metal do not warn you that you are now engaged upon a senseless task. You remind me of the Pelopaeus[2], who used to coat with mud the place on the wall whence her nest had been removed. You speak to me, in your own fashion, of a strange psychology which is able to reconcile the wonders of a master craftsmanship with aberrations due to unfathomable stupidity.

Let us compare the work of the Banded Epeira with that of the Penduline Titmouse, the cleverest of our small birds in the art of nest-building. This Tit haunts the osier-beds of the lower reaches of the Rhone. Rocking gently in the river breeze, his nest sways pendent over the peaceful backwaters, at some distance from the too-impetuous current. It hangs from the drooping end of the

[1]These experiments are described in the author's essay on the Mason Bees entitled *Fragments on Insect Psychology.*—*Translator's Note.*

[2]A species of Wasp.—*Translator's Note.*

branch of a poplar, an old willow or an alder, all of them tall trees, favouring the banks of streams.

It consists of a cotton bag, closed all round, save for a small opening at the side, just sufficient to allow of the mother's passage. In shape, it resembles the body of an alembic, a chemist's retort with a short lateral neck, or, better still, the foot of a stocking, with the edges brought together, but for a little round hole left at one side. The outward appearances increase the likeness: one can almost see the traces of a knitting-needle working with coarse stitches. That is why, struck by this shape, the Provençal peasant, in his expressive language, calls the Penduline *lou Debassaire*, the Stocking-knitter.

The early-ripening seedlets of the widows and poplars furnish the materials for the work. There breaks from them, in May, a sort of vernal snow, a fine down, which the eddies of the air heap in the crevices of the ground. It is a cotton similar to that of our manufactures, but of very short staple. It comes from an inexhaustible warehouse: the tree is bountiful; and the wind from the osier-beds gathers the tiny flocks as they pour from the seeds. They are easy to pick up.

The difficulty is to set to work. How does the bird proceed, in order to knit its stocking? How, with such simple implements as its beak and claws, does it manage to produce a fabric which our skilled fingers would fail to achieve? An examination of the nest will inform us, to a certain extent.

The cotton of the poplar cannot, of itself, supply a hanging pocket capable of supporting the weight of the brood and resisting the buffeting of the wind. Rammed, entangled and packed together, the flocks, similar to those which ordinary wadding would give if chopped up very fine, would produce only an agglomeration devoid of cohesion and liable to be dispelled by the first breath of air. They require a canvas, a warp, to keep them in position.

Tiny dead stalks, with fibrous barks, well softened by the action of moisture and the air, furnish the Penduline with a coarse tow, not unlike that of hemp. With these ligaments, purged of every woody particle and tested for flexibility and tenacity, he winds a number of loops round the end of the branch which he has selected as a support for his structure.

It is not a very accurate piece of work. The loops run clumsily and anyhow: some are slacker, others tighter; but, when all is said, it is solid, which is the main point. Also, this fibrous sheath, the keystone of the edifice, occupies a fair length of branch, which enables the fastenings for the net to be multiplied.

The several straps, after describing a certain number of turns, ravel out at the ends and hang loose. After them come interlaced threads, greater in number and finer in texture. In the tangled jumble occur what might almost be described as weaver's knots. As far as one can judge by the result alone, without having seen the bird at work, this is how the canvas, the support of the cotton wall, is obtained.

This warp, this inner framework, is obviously not constructed in its entirety from the start; it goes on gradually, as the bird stuffs the part above it with cotton. The wadding, picked up bit by bit from the ground, is teazled by the bird's claws and inserted, all fleecy, into the meshes of the canvas. The beak pushes it, the breast presses it, both inside and out. The result is a soft felt a couple of inches thick.

Near the top of the pouch, on one side, is contrived a narrow orifice, tapering into a short neck. This is the kitchen-door. In order to pass through it, the Penduline, small though he be, has to force the elastic partition, which yields slightly and then contracts. Lastly, the house is furnished with a mattress of first-quality cotton. Here lie from six to eight white eggs, the size of a cherry-stone.

Well, this wonderful nest is a barbarous casemate compared with that of the Banded Epeira. As regards shape, this stocking-foot cannot be mentioned in the same breath with the Spider's elegant and faultlessly-rounded balloon. The fabric of mixed cotton and tow is a rustic frieze beside the spinstress' satin; the suspension-straps are clumsy cables compared with her delicate silk fastenings. Where shall we find in the Penduline's mattress aught to vie with the Epeira's eiderdown, that teazled russet gossamer? The Spider is superior to the bird in every way, in so far as concerns her work.

But, on her side, the Penduline is a more devoted mother. For weeks on end, squatting at the bottom of her purse, she presses to her heart the eggs, those little white pebbles from which the warmth of her body will bring forth life. The Epeira knows not these softer passions. Without bestowing a second glance an it, she abandons her nest to its fate, be it good or ill.

CHAPTER III

THE NARBONNE LYCOSA

THE Epeira, who displays such astonishing industry to give her eggs a dwelling-house of incomparable perfection, becomes, after that, careless of her family. For what reason? She lacks the time. She has to die when the first cold comes, whereas the eggs are destined to pass the winter in their downy snuggery. The desertion of the nest is inevitable, owing to the very force of things. But, if the hatching were earlier and took place in the Epeira's lifetime, I imagine that she would rival the bird in devotion.

So I gather from the analogy of *Thomisus onustus*, WALCK., a shapely Spider who weaves no web, lies in wait for her prey and walks sideways, after the manner of the Crab. I have spoken elsewhere[1] of her encounters with the Domestic Bee, whom she jugulates by biting her in the neck.

Skilful in the prompt despatch of her prey, the little Crab Spider is no less well-versed in the nesting art. I find her settled on a privet in the enclosure. Here, in the heart of a cluster of flowers, the luxurious creature plaits a little pocket of white satin, shaped like a wee thimble. It is the receptacle for the eggs. A round, flat lid, of a felted fabric, closes the mouth.

Above this ceiling rises a dome of stretched threads and faded flowerets which have fallen from the cluster. This is the watcher's belvedere, her conning-tower. An opening, which is always free, gives access to this post.

Here the Spider remains on constant duty. She has thinned greatly since she laid her eggs, has almost lost her corporation. At the least alarm, she sallies forth, waves a threatening limb at the passing stranger and invites him, with a gesture, to keep his distance. Having put the intruder to flight, she quickly returns indoors.

[1]In Chap. VIII. of the present volume.—*Translator's Note*.

And what does she do in there, under her arch of withered flowers and silk? Night and day, she shields the precious eggs with her poor body spread out flat. Eating is neglected. No more lying in wait, no more Bees drained to the last drop of blood. Motionless, rapt in meditation, the Spider is in an incubating posture, in other words, she is sitting on her eggs. Strictly speaking, the word 'incubating' means that and nothing else.

The brooding Hen is no more assiduous, but she is also a heating-apparatus and, with the gentle warmth of her body, awakens the germs to life. For the Spider, the heat of the sun suffices; and this alone keeps me from saying that she 'broods.'

For two or three weeks, more and more wrinkled by abstinence, the little Spider never relaxes her position. Then comes the hatching. The youngsters stretch a few threads in swing-like curves from twig to twig. The tiny rope-dancers practise for some days in the sun; then they disperse, each intent upon his own affairs.

Let us now look at the watch-tower of the nest. The mother is still there, but this time lifeless. The devoted creature has known the delight of seeing her family born; she has assisted the weaklings through the trap-door; and, when her duty was done, very gently she died. The Hen does not reach this height of self-abnegation.

Other Spiders do better still, as, for instance, the Narbonne Lycosa, or Black-bellied Tarantula (*Lycosa narbonnensis*, WALCK.), whose prowess has been described in an earlier chapter. The reader will remember her burrow, her pit of a bottle-neck's width, dug in the pebbly soil beloved by the lavender and the thyme. The mouth is rimmed by a bastion of gravel and bits of wood cemented with silk. There is nothing else around her dwelling: no web, no snares of any kind.

From her inch-high turret, the Lycosa lies in wait for the passing Locust. She gives a bound, pursues the prey and suddenly deprives it of motion with a bite in the neck. The game is consumed on the spot, or else in the lair; the insect's tough hide arouses no disgust. The sturdy huntress is not a drinker of blood, like the Epeira; she needs solid food, food that crackles between the jaws. She is like a Dog devouring his bone.

Would you care to bring her to the light of day from the depths of her well? Insert a thin straw into the burrow and move it about.

Uneasy as to what is happening above, the recluse hastens to climb up and stops, in a threatening attitude, at some distance from the orifice. You see her eight eyes gleaming like diamonds in the dark; you see her powerful poison-fangs yawning, ready to bite. He who is not accustomed to the sight of this horror, rising from under the ground, cannot suppress a shiver. B-r-r-r-r! Let us leave the beast alone.

Chance, a poor stand-by, sometimes contrives very well. At the beginning of the month of August, the children call me to the far side of the enclosure, rejoicing in a find which they have made under the rosemary-bushes. It is a magnificent Lycosa, with an enormous belly, the sign of an impending delivery.

The obese Spider is gravely devouring something in the midst of a circle of onlookers. And what? The remains of a Lycosa a little smaller than herself, the remains of her male. It is the end of the tragedy that concludes the nuptials. The sweetheart is eating her lover. I allow the matrimonial rites to be fulfilled in all their horror; and, when the last morsel of the unhappy wretch has been scrunched up, I incarcerate the terrible matron under a cage standing in an earthen pan filled with sand.

Early one morning, ten days later, I find her preparing for her confinement. A silk network is first spun on the ground, covering an extent about equal to the palm of one's hand. It is coarse and shapeless, but firmly fixed. This is the floor on which the Spider means to operate.

On this foundation, which acts as a protection from the sand, the Lycosa fashions a round mat, the size of a two-franc piece and made of superb white silk. With a gentle, uniform movement, which might be regulated by the wheels of a delicate piece of clockwork, the tip of the abdomen rises and falls, each time touching the supporting base a little farther away, until the extreme scope of the mechanism is attained.

Then, without the Spider's moving her position, the oscillation is resumed in the opposite direction. By means of this alternate motion, interspersed with numerous contacts, a segment of the sheet is obtained, of a very accurate texture. When this is done, the Spider moves a little along a circular line and the loom works in the same manner on another segment.

The silk disk, a sort of hardly concave paten, now no longer receives aught from the spinnerets in its centre; the marginal belt alone increases in thickness. The piece thus becomes a bowl-shaped porringer, surrounded by a wide, flat edge.

The time for the laying has come. With one quick emission, the viscous, pale-yellow eggs are laid in the basin, where they heap together in the shape of a globe which projects largely outside the cavity. The spinnerets are once more set going. With short movements, as the tip of the abdomen rises and falls to weave the round mat, they cover up the exposed hemisphere. The result is a pill set in the middle of a circular carpet.

The legs, hitherto idle, are now working. They take up and break off one by one the threads that keep the round mat stretched on the coarse supporting network. At the same time, the fangs grip this sheet, lift it by degrees, tear it from its base and fold it over upon the globe of eggs. It is a laborious operation. The whole edifice totters, the floor collapses, fouled with sand. By a movement of the legs, those soiled shreds are cast aside. Briefly, by means of violent tugs of the fangs, which pull, and broom-like efforts of the legs, which clear away, the Lycosa extricates the bag of eggs and removes it as a clear-cut mass, free from any adhesion.

It is a white-silk pill, soft to the touch and glutinous. Its size is that of an average cherry. An observant eye will notice, running horizontally around the middle, a fold which a needle is able to raise without breaking it. This hem, generally undistinguishable from the rest of the surface, is none other than the edge of the circular mat, drawn over the lower hemisphere. The other hemisphere, through which the youngsters will go out, is less well fortified: its only wrapper is the texture spun over the eggs immediately after they were laid.

Inside, there is nothing but the eggs: no mattress, no soft eiderdown, like that of the Epeirae. The Lycosa, indeed, has no need to guard her eggs against the inclemencies of the winter, for the hatching will take place long before the cold weather comes. Similarly, the Thomisus, with her early brood, takes good care not to incur useless expenditure: she gives her eggs, for their protection, a simple purse of satin.

THE FANGS GRIP THIS SHEET... TEAR IT FROM IT'S BASE AND FOLD IT OVER UPON THE GLOBE OF EGGS (UPPER).

THE EPERIA... TURNS ON HER BACK AND ROLLS ON THE GROUND IN THE MANNER OF A DONKEY (LOWER).

The work of spinning, followed by that of tearing, is continued for a whole morning, from five to nine o'clock. Worn out with fatigue, the mother embraces her dear pill and remains motionless. I shall see no more to-day. Next morning, I find the Spider carrying the bag of eggs slung from her stern.

Henceforth, until the hatching, she does not leave go of the precious burden, which, fastened to the spinnerets by a short ligament, drags and bumps along the ground. With this load banging against her heels, she goes about her business; she walks or rests, she seeks her prey, attacks it and devours it. Should some accident cause the wallet to drop off, it is soon replaced. The spinnerets touch it somewhere, anywhere, and that is enough: adhesion is at once restored.

The Lycosa is a stay-at-home. She never goes out except to snap up some game passing within her hunting-domains, near the burrow. At the end of August, however, it is not unusual to meet her roaming about, dragging her wallet behind her. Her hesitations make one think that she is looking for her home, which she has left for the moment and has a difficulty in finding.

Why these rambles? There are two reasons: first the pairing and then the making of the pill. There is a lack of space in the burrow, which provides only room enough for the Spider engaged in long contemplation. Now the preparations for the egg-bag require an extensive flooring, a supporting framework about the size of one's hand, as my caged prisoner has shown us. The Lycosa has not so much space at her disposal, in her well; hence the necessity for coming out and working at her wallet in the open air, doubtless in the quiet hours of the night.

The meeting with the male seems likewise to demand an excursion. Running the risk of being eaten alive, will he venture to plunge into his lady's cave, into a lair whence flight would be impossible? It is very doubtful. Prudence demands that matters should take place outside. Here at least there is some chance of beating a hasty retreat which will enable the rash swain to escape the attacks of his horrible bride.

The interview in the open air lessens the danger without removing it entirely. We had proof of this when we caught the Lycosa in the act of devouring her lover aboveground, in a part of the enclosure which had been broken for planting and which was therefore not suitable for the Spider's establishment. The burrow must have been some way off; and the meeting of the pair took place at the very spot of the tragic catastrophe. Although he had a clear road, the male was not quick enough in getting away and was duly eaten.

After this cannibal orgy, does the Lycosa go back home? Perhaps not, for a while. Besides, she would have to go out a second time, to manufacture her pill on a level space of sufficient extent.

When the work is done, some of them emancipate themselves, think they will have a look at the country before retiring for good and all. It is these whom we sometimes meet wandering aimlessly and dragging their bag behind them. Sooner or later, however, the vagrants return home; and the month of August is not over before a straw rustled in any burrow will bring the mother up, with her wallet slung behind her. I am able to procure as many as I want and, with them, to indulge in certain experiments of the highest interest.

It is a sight worth seeing, that of the Lycosa dragging her treasure after her, never leaving it, day or night, sleeping or waking, and defending it with a courage that strikes the beholder with awe. If I try to take the bag from her, she presses it to her breast in despair, hangs on to my pincers, bites them with her poison-fangs. I can hear the daggers grating on the steel. No, she would not allow herself to be robbed of the wallet with impunity, if my fingers were not supplied with an implement.

By dint of pulling and shaking the pill with the forceps, I take it from the Lycosa, who protests furiously. I fling her in exchange a pill taken from another Lycosa. It is at once seized in the fangs, embraced by the legs and hung on to the spinneret. Her own or another's: it is all one to the Spider, who walks away proudly with the alien wallet. This was to be expected, in view of the similarity of the pills exchanged.

A test of another kind, with a second subject, renders the mistake more striking. I substitute, in the place of the lawful bag which I have removed, the work of the Silky Epeira. The colour and softness of the material are the same in both cases; but the shape is quite different. The stolen object is a globe; the object presented in exchange is an elliptical conoid studded with angular projections along the edge of the base. The Spider takes no account of this dissimilarity. She promptly glues the queer bag to her spinnerets and is as pleased as though she were in possession of her real pill. My experimental villainies have no other consequences beyond an ephemeral carting. When hatching-time arrives, early in the case of the Lycosa, late in that of the Epeira, the gulled Spider abandons the strange bag and pays it no further attention.

Let us penetrate yet deeper into the wallet-bearer's stupidity. After depriving the Lycosa of her eggs, I throw her a ball of cork, roughly polished with a file and of the same size as the stolen pill. She accepts the corky substance, so different from the silk purse, without the least demur. One would have thought that she would recognize her mistake with those eight eyes of hers, which gleam like precious stones. The silly creature pays no attention. Lovingly she embraces the cork ball, fondles it with her palpi, fastens it to her spinnerets and thenceforth drags it after her as though she were dragging her own bag.

Let us give another the choice between the imitation and the real. The rightful pill and the cork ball are placed together on the floor of the jar. Will the Spider be able to know the one that belongs to her? The fool is incapable of doing so. She makes a wild rush and seizes haphazard at one time her property, at another my sham product. Whatever is first touched becomes a good capture and is forthwith hung up.

If I increase the number of cork balls, if I put in four or five of them, with the real pill among them, it is seldom that the Lycosa recovers her own property. Attempts at enquiry, attempts at selection there are none. Whatever she snaps up at random she sticks to, be it good or bad. As there are more of the sham pills of cork, these are the most often seized by the Spider.

This obtuseness baffles me. Can the animal be deceived by the soft contact of the cork? I replace the cork balls by pellets of cotton or paper, kept in their round shape with a few bands of thread. Both are very readily accepted instead of the real bag that has been removed.

Can the illusion be due to the colouring, which is light in the cork and not unlike the tint of the silk globe when soiled with a little earth, while it is white in the paper and the cotton, when it is identical with that of the original pill? I give the Lycosa, in exchange for her work, a pellet of silk thread, chosen of a fine red, the brightest of all colours. The uncommon pill is as readily accepted and as jealously guarded as the others.

We will leave the wallet-bearer alone; we know all that we want to know about her poverty of intellect. Let us wait for the hatching, which takes place in the first fortnight in September. As they come out of the pill, the youngsters, to the number of about a couple of hundred, clamber on the Spider's back and there sit motionless, jammed close together, forming a sort of bark of mingled legs and paunches. The mother is unrecognizable under this live mantilla. When the hatching is over, the wallet is loosened from the spinnerets and cast aside as a worthless rag.

The little ones are very good: none stirs none tries to get more room for himself at his neighbours' expense. What are they doing there, so quietly? They allow themselves to be carted about, like the young of the Opossum. Whether she sit in long meditation at the bottom of her den, or come to the orifice, in mild weather, to bask in the sun, the Lycosa never throws off her great-coat of swarming youngsters until the fine season comes.

If, in the middle of winter, in January or February, I happen, out in the fields, to ransack the Spider's dwelling, after the rain, snow and frost have battered it and, as a rule, dismantled the bastion at the entrance, I always find her at home, still full of vigour, still carrying her family. This vehicular upbringing lasts five or six months at least, without interruption. The celebrated American carrier, the Opossum, who emancipates her offspring after a few weeks' carting, cuts a poor figure beside the Lycosa.

What do the little ones eat, on the maternal spine? Nothing, so far as I know. I do not see them grow larger. I find them, at the tardy period of their emancipation, just as they were when they left the bag.

During the bad season, the mother herself is extremely abstemious. At long intervals, she accepts, in my jars, a belated Locust, whom I have captured, for her benefit, in the sunnier nooks. In order to keep herself in condition, as when she is dug up in the course of my winter excavations, she must therefore sometimes break her fast and come out in search of prey, without, of course, discarding her live mantilla.

The expedition has its dangers. The youngsters may be brushed off by a blade of grass. What becomes of them when they have a fall? Does the mother give them a thought? Does she come to their assistance and help them to regain their place on her back? Not at all. The affection of a Spider's heart, divided among some hundreds, can spare but a very feeble portion to each. The Lycosa hardly troubles, whether one youngster fall from his place, or six, or all of them. She waits impassively for the victims of the mishap to get out of their own difficulty, which they do, for that matter, and very nimbly.

I sweep the whole family from the back of one of my boarders with a hair-pencil. Not a sign of emotion, not an attempt at search on the part of the denuded one. After trotting about a little on the sand, the dislodged youngsters find, these here, those there, one or other of the mother's legs, spread wide in a circle. By means of these climbing-poles, they swarm to the top and soon the dorsal group resumes its original form. Not one of the lot is missing. The Lycosa's sons know their trade as acrobats to perfection: the mother need not trouble her head about their fall.

With a sweep of the pencil, I make the family of one Spider fall around another laden with her own family. The dislodged ones nimbly scramble up the legs and climb on the back of their new mother, who kindly allows them to behave as though they belonged to her. There is no room on the abdomen, the regulation resting-place, which is already occupied by the real sons. The invaders thereupon encamp on the front part, beset the thorax and change the carrier into a horrible pin-cushion that no longer bears the least resemblance to a Spider form. Meanwhile, the sufferer raises no sort of protest against this access of family. She placidly accepts them all and walks them all about.

The youngsters, on their side, are unable to distinguish between what is permitted and forbidden. Remarkable acrobats that they are, they climb on the first Spider that comes along, even when of a different species, provided that she be of a fair size. I place them in the presence of a big Epeira marked with a white cross on a pale-orange ground (*Epeira pallida*, OLIV.). The little ones, as soon as they are dislodged from the back of the Lycosa their mother, clamber up the stranger without hesitation.

Intolerant of these familiarities, the Spider shakes the leg encroached upon and flings the intruders to a distance. The assault is doggedly resumed, to such good purpose that a dozen succeed in hoisting themselves to the top. The Epeira, who is not accustomed to the tickling of such a load, turns over on her back and rolls on the ground in the manner of a donkey when his hide is itching. Some are lamed, some are even crushed. This does not deter the others, who repeat the escalade as soon as the Epeira is on her legs again. Then come more somersaults, more rollings on the back, until the giddy swarm are all discomfited and leave the Spider in peace.

CHAPTER IV

THE NARBONNE LYCOSA: THE BURROW

MICHELET[1] has told us how, as a printer's apprentice in a cellar, he established amicable relations with a Spider. At a certain hour of the day, a ray of sunlight would glint through the window of the gloomy workshop and light up the little compositor's case. Then his eight-legged neighbour would come down from her web and take her share of the sunshine on the edge of the case. The boy did not interfere with her; he welcomed the trusting visitor as a friend and as a pleasant diversion from the long monotony. When we lack the society of our fellow-men, we take refuge in that of animals, without always losing by the change.

I do not, thank God, suffer from the melancholy of a cellar: my solitude is gay with light and verdure; I attend, whenever I please, the fields' high festival, the Thrushes' concert, the Crickets' symphony; and yet my friendly commerce with the Spider is marked by an even greater devotion than the young typesetter's. I admit her to the intimacy of my study, I make room for her among my books, I set her in the sun on my window-ledge, I visit her assiduously at her home, in the country. The object of our relations is not to create a means of escape from the petty worries of life, pin-pricks whereof I have my share like other men, a very large share, indeed; I propose to submit to the Spider a host of questions whereto, at times, she condescends to reply.

To what fair problems does not the habit of frequenting her give rise! To set them forth worthily, the marvellous art which the little printer was to acquire were not too much. One needs the pen of a

---

[1] Jules Michelet (1798-1874), author of L'Oiseau and L'Insecte, in addition to the historical works for which he is chiefly known. As a lad, he helped his father, a printer by trade, in setting type.—*Translator's Note*.

Michelet; and I have but a rough, blunt pencil. Let us try, nevertheless: even when poorly clad, truth is still beautiful.

I will therefore once more take up the story of the Spider's instinct, a story of which the preceding chapters have given but a very rough idea. Since I wrote those earlier essays, my field of observation has been greatly extended. My notes have been enriched by new and most remarkable facts. It is right that I should employ them for the purpose of a more detailed biography.

The exigencies of order and clearness expose me, it is true, to occasional repetitions. This is inevitable when one has to marshal in an harmonious whole a thousand items culled from day to day, often unexpectedly, and bearing no relation one to the other. The observer is not master of his time; opportunity leads him and by unsuspected ways. A certain question suggested by an earlier fact finds no reply until many years after. Its scope, moreover, is amplified and completed with views collected on the road. In a work, therefore, of this fragmentary character, repetitions, necessary for the due co-ordination of ideas, are inevitable. I shall be as sparing of them as I can.

Let us once more introduce our old friends the Epeira and the Lycosa, who are the most important Spiders in my district. The Narbonne Lycosa, or Black-bellied Tarantula, chooses her domicile in the waste, pebbly lands beloved of the thyme. Her dwelling, a fortress rather than a villa, is a burrow about nine inches deep and as wide as the neck of a claret-bottle. The direction is perpendicular, in so far as obstacles, frequent in a soil of this kind, permit. A bit of gravel can be extracted and hoisted outside; but a flint is an immovable boulder which the Spider avoids by giving a bend to her gallery. If more such are met with, the residence becomes a winding cave, with stone vaults, with lobbies communicating by means of sharp passages.

This lack of plan has no attendant drawbacks, so well does the owner, from long habit, know every corner and storey of her mansion. If any interesting buzz occur overhead, the Lycosa climbs up from her rugged manor with the same speed as from a vertical shaft. Perhaps she even finds the windings and turnings an advantage, when she has to drag into her den a prey that happens to defend itself.

As a rule, the end of the burrow widens into a side-chamber, a lounge or resting-place where the Spider meditates at length and is content to lead a life of quiet when her belly is full.

A silk coating, but a scanty one, for the Lycosa has not the wealth of silk possessed by the Weaving Spiders, lines the walls of the tube and keeps the loose earth from falling. This plaster, which cements the incohesive and smooths the rugged parts, is reserved more particularly for the top of the gallery, near the mouth. Here, in the daytime, if things be peaceful all around, the Lycosa stations herself, either to enjoy the warmth of the sun, her great delight, or to lie in wait for game. The threads of the silk lining afford a firm hold to the claws on every side, whether the object be to sit motionless for hours, revelling in the light and heat, or to pounce upon the passing prey.

Around the orifice of the burrow rises, to a greater or lesser height, a circular parapet, formed of tiny pebbles, twigs and straps borrowed from the dry leaves of the neighbouring grasses, all more or less dexterously tied together and cemented with silk. This work of rustic architecture is never missing, even though it be no more than a mere pad.

When she reaches maturity and is once settled, the Lycosa becomes eminently domesticated. I have been living in close communion with her for the last three years. I have installed her in large earthen pans on the window-sills of my study and I have her daily under my eyes. Well, it is very rarely that I happen on her outside, a few inches from her hole, back to which she bolts at the least alarm.

We may take it, then, that, when not in captivity, the Lycosa does not go far afield to gather the wherewithal to build her parapet and that she makes shift with what she finds upon her threshold. In these conditions, the building-stones are soon exhausted and the masonry ceases for lack of materials.

The wish came over me to see what dimensions the circular edifice would assume, if the Spider were given an unlimited supply. With captives to whom I myself act as purveyor the thing is easy enough. Were it only with a view to helping whoso may one day care to continue these relations with the big Spider of the wastelands, let me describe how my subjects are housed.

A good-sized earthenware pan, some nine inches deep, is filled with a red, clayey earth, rich in pebbles, similar, in short, to that of the places haunted by the Lycosa. Properly moistened into a paste, the artificial soil is heaped, layer by layer, around a central reed, of a bore equal to that of the animal's natural burrow. When the receptacle is filled to the top, I withdraw the reed, which leaves a yawning, perpendicular shaft. I thus obtain the abode which shall replace that of the fields.

To find the hermit to inhabit it is merely the matter of a walk in the neighbourhood. When removed from her own dwelling, which is turned topsy-turvy by my trowel, and placed in possession of the den produced by my art, the Lycosa at once disappears into that den. She does not come out again, seeks nothing better elsewhere. A large wire-gauze cover rests on the soil in the pan and prevents escape.

In any case, the watch, in this respect, makes no demands upon my diligence. The prisoner is satisfied with her new abode and manifests no regret for her natural burrow. There is no attempt at flight on her part. Let me not omit to add that each pan must receive not more than one inhabitant. The Lycosa is very intolerant. To her, a neighbour is fair game, to be eaten without scruple when one has might on one's side. Time was when, unaware of this fierce intolerance, which is more savage still at breeding-time, I saw hideous orgies perpetrated in my overstocked cages. I shall have occasion to describe those tragedies later.

Let us meanwhile consider the isolated Lycosae. They do not touch up the dwelling which I have moulded for them with a bit of reed; at most, now and again, perhaps with the object of forming a lounge or bedroom at the bottom, they fling out a few loads of rubbish. But all, little by little, build the kerb that is to edge the mouth.

I have given them plenty of first-rate materials, far superior to those which they use when left to their own resources. These consist, first, for the foundations, of little smooth stones, some of which are as large as an almond. With this road-metal are mingled short strips of raphia, or palm-fibre, flexible ribbons, easily bent. These stand for the Spider's usual basket-work, consisting of slender stalks and dry blades of grass. Lastly, by way of an unprecedented treasure, never yet employed by a Lycosa, I place at my captives' disposal some thick threads of wool, cut into inch lengths.

As I wish, at the same time, to find out whether my animals, with the magnificent lenses of their eyes, are able to distinguish colours and prefer one colour to another, I mix up bits of wool of different hues: there are red, green, white and yellow pieces. If the Spider have any preference, she can choose where she pleases.

The Lycosa always works at night, a regrettable circumstance, which does not allow me to follow the worker's methods. I see the result; and that is all. Were I to visit the building-yard by the light of a lantern, I should be no wiser. The animal, which is very shy, would at once dive into her lair; and I should have lost my sleep for nothing. Furthermore, she is not a very diligent labourer; she likes to take her time. Two or three bits of wool or raphia placed in position represent a whole night's work. And to this slowness we must add long spells of utter idleness.

Two months pass; and the result of my liberality surpasses my expectations. Possessing more windfalls than they know what to do with, all picked up in their immediate neighbourhood, my Lycosae have built themselves donjon-keeps the like of which their race has not yet known. Around the orifice, on a slightly sloping bank, small, flat, smooth stones have been laid to form a broken, flagged pavement. The larger stones, which are Cyclopean blocks compared with the size of the animal that has shifted them, are employed as abundantly as the others.

On this rockwork stands the donjon. It is an interlacing of raphia and bits of wool, picked up at random, without distinction of shade. Red and white, green and yellow are mixed without any attempt at order. The Lycosa is indifferent to the joys of colour.

The ultimate result is a sort of muff, a couple of inches high. Bands of silk, supplied by the spinnerets, unite the pieces, so that the whole resembles a coarse fabric. Without being absolutely faultless, for there are always awkward pieces on the outside, which the worker could not handle, the gaudy building is not devoid of merit. The bird lining its nest would do no better. Whoso sees the curious, many-coloured productions in my pans takes them for an outcome of my industry, contrived with a view to some experimental mischief; and his surprise is great when I confess who the real author is. No one would ever believe the Spider capable of constructing such a monument.

It goes without saying that, in a state of liberty, on our barren waste-lands, the Lycosa does not indulge in such sumptuous architecture. I have given the reason: she is too great a stay-at-home to go in search of materials and she makes use of the limited resources which she finds around her. Bits of earth, small chips of stone, a few twigs, a few withered grasses: that is all, or nearly all. Wherefore the work is generally quite modest and reduced to a parapet that hardly attracts attention.

My captives teach us that, when materials are plentiful, especially textile materials that remove all fears of landslip, the Lycosa delights in tall turrets. She understands the art of donjon-building and puts it into practice as often as she possesses the means.

This art is akin to another, from which it is apparently derived. If the sun be fierce or if rain threaten, the Lycosa closes the entrance to her dwelling with a silken trellis-work, wherein she embeds different matters, often the remnants of victims which she has devoured. The ancient Gael nailed the heads of his vanquished enemies to the door of his hut. In the same way, the fierce Spider sticks the skulls of her prey into the lid of her cave. These lumps look very well on the ogre's roof; but we must be careful not to mistake them for warlike trophies. The animal knows nothing of our barbarous bravado. Everything at the threshold of the burrow is used indiscriminately: fragments of Locust, vegetable remains and especially particles of earth. A Dragon-fly's head baked by the sun is as good as a bit of gravel and no better.

And so, with silk and all sorts of tiny materials, the Lycosa builds a lidded cap to the entrance of her home. I am not well acquainted with the reasons that prompt her to barricade herself indoors, particularly as the seclusion is only temporary and varies greatly in duration. I obtain precise details from a tribe of Lycosae wherewith the enclosure, as will be seen later, happens to be thronged in consequence of my investigations into the dispersal of the family.

At the time of the tropical August heat, I see my Lycosae, now this batch, now that, building, at the entrance to the burrow, a convex ceiling, which is difficult to distinguish from the surrounding soil. Can it be to protect themselves from the too-vivid light? This is doubtful; for, a few days later, though the power of the sun remain the same, the roof is broken open and the Spider reappears at her door, where she revels in the torrid heat of the dog-days.

Later, when October comes, if it be rainy weather, she retires once more under a roof, as though she were guarding herself against the damp. Let us not be too positive of anything, however: often, when it is raining hard, the Spider bursts her ceiling and leaves her house open to the skies.

Perhaps the lid is only put on for serious domestic events, notably for the laying. I do, in fact, perceive young Lycosae who shut themselves in before they have attained the dignity of motherhood and who reappear, some time later, with the bag containing the eggs hung to their stern. The inference that they close the door with the object of securing greater quiet while spinning the maternal cocoon would not be in keeping with the unconcern displayed by the majority. I find some who lay their eggs in an open burrow; I come upon some who weave their cocoon and cram it with eggs in the open air, before they even own a residence. In short, I do not succeed in fathoming the reasons that cause the burrow to be closed, no matter what the weather, hot or cold, wet or dry.

The fact remains that the lid is broken and repaired repeatedly, sometimes on the same day. In spite of the earthy casing, the silk woof gives it the requisite pliancy to cleave when pushed by the anchorite and to rip open without falling into ruins. Swept back to the circumference of the mouth and increased by the wreckage of further ceilings, it becomes a parapet, which the Lycosa raises by degrees in her long moments of leisure. The bastion which surmounts the burrow, therefore, takes its origin from the temporary lid. The turret derives from the split ceiling.

What is the purpose of this turret? My pans will tell us that. An enthusiastic votary of the chase, so long as she is not permanently fixed, the Lycosa, once she has set up house, prefers to lie in ambush and wait for the quarry. Every day, when the heat is greatest, I see my captives come up slowly from under ground and lean upon the battlements of their woolly castle-keep. They are then really magnificent in their stately gravity. With their swelling belly contained within the aperture, their head outside, their glassy eyes staring, their legs gathered for a spring, for hours and hours they wait, motionless, bathing voluptuously in the sun.

Should a tit-bit to her liking happen to pass, forthwith the watcher darts from her tall tower, swift as an arrow from the bow. With a dagger-thrust in the neck, she stabs the jugular of the Locust, Dragon-fly or other prey whereof I am the purveyor; and she as quickly scales the donjon and retires with her capture. The performance is a wonderful exhibition of skill and speed.

Very seldom is a quarry missed, provided that it pass at a convenient distance, within the range of the huntress' bound. But, if the prey be at some distance, for instance on the wire of the cage, the Lycosa takes no notice of it. Scorning to go in pursuit, she allows it to roam at will. She never strikes except when sure of her stroke. She achieves this by means of her tower. Hiding behind the wall, she sees the stranger advancing, keeps her eyes on him and suddenly pounces when he comes within reach. These abrupt tactics make the thing a certainty. Though he were winged and swift of flight, the unwary one who approaches the ambush is lost.

This presumes, it is true, an exemplary patience on the Lycosa's part; for the burrow has naught that can serve to entice victims. At best, the ledge provided by the turret may, at rare intervals, tempt some weary wayfarer to use it as a resting-place. But, if the quarry do not come to-day, it is sure to come to-morrow, the next day, or later, for the Locusts hop innumerable in the waste-land, nor are they always able to regulate their leaps. Some day or other, chance is bound to bring one of them within the purlieus of the burrow. This is the moment to spring upon the pilgrim from the ramparts. Until then, we maintain a stoical vigilance. We shall dine when we can; but we shall end by dining.

The Lycosa, therefore, well aware of these lingering eventualities, waits and is not unduly distressed by a prolonged abstinence. She has an accommodating stomach, which is satisfied to be gorged to-day and to remain empty afterwards for goodness knows how long. I have sometimes neglected my catering-duties for weeks at a time; and my boarders have been none the worse for it. After a more or less protracted fast, they do not pine away, but are smitten with a wolf-like hunger. All these ravenous eaters are alike: they guzzle to excess to-day, in anticipation of to-morrow's dearth.

In her youth, before she has a burrow, the Lycosa earns her living in another manner. Clad in grey like her elders, but without the black-velvet apron which she receives on attaining the marriageable age, she roams among the scrubby grass. This is true hunting. Should a suitable quarry heave in sight, the Spider pursues it, drives it from its shelters, follows it hot-foot. The fugitive gains the heights, makes as though to fly away. He has not the time. With an upward leap, the Lycosa grabs him before he can rise.

I am charmed with the agility wherewith my yearling boarders seize the Flies which I provide for them. In vain does the Fly take refuge a couple of inches up, on some blade of grass. With a sudden spring into the air, the Spider pounces on the prey. No Cat is quicker in catching her Mouse.

But these are the feats of youth not handicapped by obesity. Later, when a heavy paunch, dilated with eggs and silk, has to be trailed along, those gymnastic performances become impracticable. The Lycosa then digs herself a settled abode, a hunting-box, and sits in her watch-tower, on the look-out for game.

When and how is the burrow obtained wherein the Lycosa, once a vagrant, now a stay-at-home, is to spend the remainder of her long life? We are in autumn, the weather is already turning cool. This is how the Field Cricket sets to work: as long as the days are fine and the nights not too cold, the future chorister of spring rambles over the fallows, careless of a local habitation. At critical moments, the cover of a dead leaf provides him with a temporary shelter. In the end, the burrow, the permanent dwelling, is dug as the inclement season draws nigh.

The Lycosa shares the Cricket's views: like him, she finds a thousand pleasures in the vagabond life. With September comes the nuptial badge, the black-velvet bib. The Spiders meet at night, by the soft moonlight: they romp together, they eat the beloved shortly after the wedding; by day, they scour the country, they track the game on the short-pile, grassy carpet, they take their fill of the joys of the sun. That is much better than solitary meditation at the bottom of a well. And so it is not rare to see young mothers dragging

THE SPIDERS MEET AT NIGHT... THEY ROMP TOGETHER, THEY EAT THE BELOVED SHORTLY AFTER THE WEDDING

their bag of eggs, or even already carrying their family, and as yet without a home.

In October, it is time to settle down. We then, in fact, find two sorts of burrows, which differ in diameter. The larger, bottle-neck burrows belong to the old matrons, who have owned their house for two years at least. The smaller, of the width of a thick lead-pencil, contain the young mothers, born that year. By dint of long and leisurely alterations, the novice's earths will increase in depth as well as in diameter and become roomy abodes, similar to those of the grandmothers. In both, we find the owner and her family, the latter sometimes already hatched and sometimes still enclosed in the satin wallet.

Seeing no digging-tools, such as the excavation of the dwelling seemed to me to require, I wondered whether the Lycosa might not avail herself of some chance gallery, the work of the Cicada or the Earth-worm. This ready-made tunnel, thought I, must shorten the labours of the Spider, who appears to be so badly off for tools; she would only have to enlarge it and put it in order. I was wrong: the burrow is excavated, from start to finish, by her unaided labour.

Then where are the digging-implements? We think of the legs, of the claws. We think of them, but reflection tells us that tools such as these would not do: they are too long and too difficult to wield in a confined space. What is required is the miner's short-handled pick, wherewith to drive hard, to insert, to lever and to extract; what is required is the sharp point that enters the earth and crumbles it into fragments. There remain the Lycosa's fangs, delicate weapons which we at first hesitate to associate with such work, so illogical does it seem to dig a pit with surgeon's scalpels.

The fangs are a pair of sharp, curved points, which, when at rest, crook like a finger and take shelter between two strong pillars. The Cat sheathes her claws under the velvet of the paw, to preserve their edge and sharpness. In the same way, the Lycosa protects her poisoned daggers by folding them within the case of two powerful columns, which come plumb on the surface and contain the muscles that work them.

Well, this surgical outfit, intended for stabbing the jugular artery of the prey, suddenly becomes a pick-axe and does rough navvy's work. To witness the underground digging is impossible; but we can, at least, with the exercise of a little patience, see the rubbish carted away. If I watch my captives, without tiring, at a very early hour—for the work takes place mostly at night and at long intervals—in the end I catch them coming up with a load. Contrary to what I expected, the legs take no part in the carting. It is the mouth that acts as the barrow. A tiny ball of earth is held between the fangs and is supported by the palpi, or feelers, which are little arms employed in the service of the mouth-parts. The Lycosa descends cautiously from her turret, goes to some distance to get rid of her burden and quickly dives down again to bring up more.

We have seen enough: we know that the Lycosa's fangs, those lethal weapons, are not afraid to bite into clay and gravel. They knead the excavated rubbish into pellets, take up the mass of earth and carry it outside. The rest follows naturally; it is the fangs that dig, delve and extract. How finely-tempered they must be, not to be blunted by this well-sinker's work and to do duty presently in the surgical operation of stabbing the neck!

I have said that the repairs and extensions of the burrow are made at long intervals. From time to time, the circular parapet receives additions and becomes a little higher; less frequently still, the dwelling is enlarged and deepened. As a rule, the mansion remains as it was for a whole season. Towards the end of winter, in March more than at any other period, the Lycosa seems to wish to give herself a little more space. This is the moment to subject her to certain tests.

We know that the Field Cricket, when removed from his burrow and caged under conditions that would allow him to dig himself a new home should the fit seize him, prefers to tramp from one casual shelter to another, or rather abandons every idea of creating a permanent residence. There is a short season whereat the instinct for building a subterranean gallery is imperatively aroused. When this season is past, the excavating artist, if accidentally deprived of his abode, becomes a wandering Bohemian, careless of a lodging. He has forgotten his talents and he sleeps out.

That the bird, the nest-builder, should neglect its art when it has no brood to care for is perfectly logical: it builds for its family, not for itself. But what shall we say of the Cricket, who is exposed to a thousand mishaps when away from home? The protection of a roof would be of great use to him; and the giddy-pate does not give it a thought, though he is very strong and more capable than ever of digging with his powerful jaws.

What reason can we allege for this neglect? None, unless it be that the season of strenuous burrowing is past. The instincts have a calendar of their own. At the given hour, suddenly they awaken; as suddenly, afterwards, they fall asleep. The ingenious become incompetent when the prescribed period is ended.

On a subject of this kind, we can consult the Spider of the wastelands. I catch an old Lycosa in the fields and house her, that same day, under wire, in a burrow where I have prepared a soil to her liking. If, by my contrivances and with a bit of reed, I have previously moulded a burrow roughly representing the one from which I took her, the Spider enters it forthwith and seems pleased with her new residence. The product of my art is accepted as her lawful property and undergoes hardly any alterations. In course of time, a bastion is erected around the orifice; the top of the gallery is cemented with silk; and that is all. In this establishment of my building, the animal's behaviour remains what it would be under natural conditions.

But place the Lycosa on the surface of the ground, without first shaping a burrow. What will the homeless Spider do? Dig herself a dwelling, one would think. She has the strength to do so; she is in the prime of life. Besides, the soil is similar to that whence I ousted her and suits the operation perfectly. We therefore expect to see the Spider settled before long in a shaft of her own construction.

We are disappointed. Weeks pass and not an effort is made, not one. Demoralized by the absence of an ambush, the Lycosa hardly vouchsafes a glance at the game which I serve up. The Crickets pass within her reach in vain; most often she scorns them. She slowly wastes away with fasting and boredom. At length, she dies.

Take up your miner's trade again, poor fool! Make yourself a home, since you know how to, and life will be sweet to you for many a long day yet: the weather is fine and victuals plentiful. Dig, delve, go underground, where safety lies. Like an idiot, you refrain; and you perish. Why?

Because the craft which you were wont to ply is forgotten; because the days of patient digging are past and your poor brain is unable to work back. To do a second time what has been done already is beyond your wit. For all your meditative air, you cannot solve the problem of how to reconstruct that which is vanished and gone.

Let us now see what we can do with younger Lycosae, who are at the burrowing-stage. I dig out five or six at the end of February. They are half the size of the old ones; their burrows are equal in diameter to my little finger. Rubbish quite fresh-spread around the pit bears witness to the recent date of the excavations.

Relegated to their wire cages, these young Lycosae behave differently according as the soil placed at their disposal is or is not already provided with a burrow made by me. A burrow is hardly the word: I give them but the nucleus of a shaft, about an inch deep, to lure them on. When in possession of this rudimentary lair, the Spider does not hesitate to pursue the work which I have interrupted in the fields. At night, she digs with a will. I can see this by the heap of rubbish flung aside. She at last obtains a house to suit her, a house surmounted by the usual turret.

The others, on the contrary, those Spiders for whom the thrust of my pencil has not contrived an entrance-hall representing, to a certain extent, the natural gallery whence I dislodged them, absolutely refuse to work; and they die, notwithstanding the abundance of provisions.

The first pursue the season's task. They were digging when I caught them; and, carried away by the enthusiasm of their activity, they go on digging inside my cages. Taken in by my decoy-shaft, they deepen the imprint of the pencil as though they were deepening their real vestibule. They do not begin their labours over again; they continue them.

The second, not having this inducement, this semblance of a burrow mistaken for their own work, forsake the idea of digging and allow themselves to die, because they would have to travel back along the chain of actions and to resume the pick-strokes of the start. To begin all over again requires reflection, a quality wherewith they are not endowed.

To the insect—and we have seen this in many earlier cases—what is done is done and cannot be taken up again. The hands of a watch do not move backwards. The insect behaves in much the same way.

Its activity urges it in one direction, ever forwards, without allowing it to retrace its steps, even when an accident makes this necessary.

What the Mason-bees and the others taught us erewhile the Lycosa now confirms in her manner. Incapable of taking fresh pains to build herself a second dwelling, when the first is done for, she will go on the tramp, she will break into a neighbour's house, she will run the risk of being eaten should she not prove the stronger, but she will never think of making herself a home by starting afresh.

What a strange intellect is that of the animal, a mixture of mechanical routine and subtle brain-power! Does it contain gleams that contrive, wishes that pursue a definite object? Following in the wake of so many others, the Lycosa warrants us in entertaining a doubt.

CHAPTER V

THE NARBONNE LYCOSA: THE FAMILY

For three weeks and more, the Lycosa trails the bag of eggs hanging to her spinnerets. The reader will remember the experiments described in the third chapter of this volume, particularly those with the cork ball and the thread pellet which the Spider so foolishly accepts in exchange for the real pill. Well, this exceedingly dull-witted mother, satisfied with aught that knocks against her heels, is about to make us wonder at her devotion.

Whether she come up from her shaft to lean upon the kerb and bask in the sun, whether she suddenly retire underground in the face of danger, or whether she be roaming the country before settling down, never does she let go her precious bag, that very cumbrous burden in walking, climbing or leaping. If, by some accident, it become detached from the fastening to which it is hung, she flings herself madly on her treasure and lovingly embraces it, ready to bite whoso would take it from her. I myself am sometimes the thief. I then hear the points of the poison-fangs grinding against the steel of my pincers, which tug in one direction while the Lycosa tugs in the other. But let us leave the animal alone: with a quick touch of the spinnerets, the pill is restored to its place; and the Spider strides off, still menacing.

Towards the end of summer, all the householders, old or young, whether in captivity on the window-sill or at liberty in the paths of the enclosure, supply me daily with the following improving sight. In the morning, as soon as the sun is hot and beats upon their burrow, the anchorites come up from the bottom with their bag and station themselves at the opening. Long siestas on the threshold in the sun are the order of the day throughout the fine season; but, at the present time, the position adopted is a different one. Formerly, the Lycosa came out into the sun for her own sake.

Leaning on the parapet, she had the front half of her body outside the pit and the hinder half inside.

The eyes took their fill of light; the belly remained in the dark. When carrying her egg-bag, the Spider reverses the posture: the front is in the pit, the rear outside. With her hind-legs she holds the white pill bulging with germs lifted above the entrance; gently she turns and returns it, so as to present every side to the life-giving rays. And this goes on for half the day, so long as the temperature is high; and it is repeated daily, with exquisite patience, during three or four weeks. To hatch its eggs, the bird covers them with the quilt of its breast; it strains them to the furnace of its heart. The Lycosa turns hers in front of the hearth of hearths, she gives them the sun as an incubator.

In the early days of September, the young ones, who have been some time hatched, are ready to come out. The pill rips open along the middle fold. We read of the origin of this fold in an earlier chapter[1]. Does the mother, feeling the brood quicken inside the satin wrapper, herself break open the vessel at the opportune moment? It seems probable. On the other hand, there may be a spontaneous bursting, such as we shall see later in the Banded Epeira's balloon, a tough wallet which opens a breach of its own accord, long after the mother has ceased to exist.

The whole family emerges from the bag straightway. Then and there, the youngsters climb to the mother's back. As for the empty bag, now a worthless shred, it is flung out of the burrow; the Lycosa does not give it a further thought. Huddled together, sometimes in two or three layers, according to their number, the little ones cover the whole back of the mother, who, for seven or eight months to come, will carry her family night and day. Nowhere can we hope to see a more edifying domestic picture than that of the Lycosa clothed in her young.

From time to time, I meet a little band of gipsies passing along the high-road on their way to some neighbouring fair. The newborn babe mewls on the mother's breast, in a hammock formed out of a kerchief. The last-weaned is carried pick-a-back; a third toddles clinging to its mother's skirts; others follow closely, the biggest in

---

[1]Chapter III. of the present volume.—*Translator's Note*.

the rear, ferreting in the blackberry-laden hedgerows. It is a magnificent spectacle of happy-go-lucky fruitfulness. They go their way, penniless and rejoicing. The sun is hot and the earth is fertile.

But how this picture pales before that of the Lycosa, that incomparable gipsy whose brats are numbered by the hundred! And one and all of them, from September to April, without a moment's respite, find room upon the patient creature's back, where they are content to lead a tranquil life and to be carted about.

The little ones are very good; none moves, none seeks a quarrel with his neighbours. Clinging together, they form a continuous drapery, a shaggy ulster under which the mother becomes unrecognizable. Is it an animal, a fluff of wool, a cluster of small seeds fastened to one another? 'Tis impossible to tell at the first glance.

The equilibrium of this living blanket is not so firm but that falls often occur, especially when the mother climbs from indoors and comes to the threshold to let the little ones take the sun. The least brush against the gallery unseats a part of the family. The mishap is not serious. The Hen, fidgeting about her Chicks, looks for the strays, calls them, gathers them together. The Lycosa knows not these maternal alarms. Impassively, she leaves those who drop off to manage their own difficulty, which they do with wonderful quickness. Commend me to those youngsters for getting up without whining, dusting themselves and resuming their seat in the saddle! The unhorsed ones promptly find a leg of the mother, the usual climbing-pole; they swarm up it as fast as they can and recover their places on the bearer's back. The living bark of animals is reconstructed in the twinkling of an eye.

To speak here of mother-love were, I think, extravagant. The Lycosa's affection for her offspring hardly surpasses that of the plant, which is unacquainted with any tender feeling and nevertheless bestows the nicest and most delicate care upon its seeds. The animal, in many cases, knows no other sense of motherhood. What cares the Lycosa for her brood! She accepts another's as readily as her own; she is satisfied so long as her back is burdened with a swarming crowd, whether it issue from her ovaries or elsewhence. There is no question here of real maternal affection.

I have described elsewhere the prowess of the Copris[1] watching over cells that are not her handiwork and do not contain her offspring. With a zeal which even the additional labour laid upon her does not easily weary, she removes the mildew from the alien dung-balls, which far exceed the regular nests in number; she gently scrapes and polishes and repairs them; she listens to them attentively and enquires by ear into each nursling's progress. Her real collection could not receive greater care. Her own family or another's: it is all one to her.

The Lycosa is equally indifferent. I take a hair-pencil and sweep the living burden from one of my Spiders, making it fall close to another covered with her little ones. The evicted youngsters scamper about, find the new mother's legs outspread, nimbly clamber up these and mount on the back of the obliging creature, who quietly lets them have their way.

They slip in among the others, or, when the layer is too thick, push to the front and pass from the abdomen to the thorax and even to the head, though leaving the region of the eyes uncovered. It does not do to blind the bearer: the common safety demands that. They know this and respect the lenses of the eyes, however populous the assembly be. The whole animal is now covered with a swarming carpet of young, all except the legs, which must preserve their freedom of action, and the under part of the body, where contact with the ground is to be feared.

My pencil forces a third family upon the already overburdened Spider; and this too is peacefully accepted. The youngsters huddle up closer, lie one on top of the other in layers and room is found for all. The Lycosa has lost the last semblance of an animal, has become a nameless bristling thing that walks about. Falls are frequent and are followed by continual climbings.

I perceive that I have reached the limits not of the bearer's goodwill, but of equilibrium. The Spider would adopt an indefinite further number of foundlings, if the dimensions of her back afforded them a firm hold. Let us be content with this. Let us restore each family to its mother, drawing at random from the lot.

---

[1] A species of Dung-beetle. Cf. *The Life and Love of the Insect*, by J. Henri Fabre, translated by Alexander Teixeira de Mattos: Chap. V.—*Translator's Note*.

There must necessarily be interchanges, but that is of no importance: real children and adopted children are the same thing in the Lycosa's eyes.

One would like to know if, apart from my artifices, in circumstances where I do not interfere, the good-natured dry-nurse sometimes burdens herself with a supplementary family; it would also be interesting to learn what comes of this association of lawful offspring and strangers. I have ample materials wherewith to obtain an answer to both questions. I have housed in the same cage two elderly matrons laden with youngsters. Each has her home as far removed from the other's as the size of the common pan permits. The distance is nine inches or more. It is not enough. Proximity soon kindles fierce jealousies between those intolerant creatures, who are obliged to live far apart, so as to secure adequate hunting-grounds.

One morning, I catch the two harridans fighting out their quarrel on the floor. The loser is laid flat upon her back; the victress, belly to belly with her adversary, clutches her with her legs and prevents her from moving a limb. Both have their poison-fangs wide open, ready to bite without yet daring, so mutually formidable are they. After a certain period of waiting, during which the pair merely exchange threats, the stronger of the two, the one on top, closes her lethal engine and grinds the head of the prostrate foe. Then she calmly devours the deceased by small mouthfuls.

Now what do the youngsters do, while their mother is being eaten? Easily consoled, heedless of the atrocious scene, they climb on the conqueror's back and quietly take their places among the lawful family. The ogress raises no objection, accepts them as her own. She makes a meal off the mother and adopts the orphans.

Let us add that, for many months yet, until the final emancipation comes, she will carry them without drawing any distinction between them and her own young. Henceforth, the two families, united in so tragic a fashion, will form but one. We see how greatly out of place it would be to speak, in this connection, of mother-love and its fond manifestations.

Does the Lycosa at least feed the younglings who, for seven months, swarm upon her back? Does she invite them to the banquet when she has secured a prize? I thought so at first; and,

SHE MAKES A MEAL OFF THE MOTHER AND ADOPTS THE ORPHANS (UPPER)

LONG SIESTAS ON THE THRESHOLD IN THE SUN ARE THE ORDER OF THE DAY (LOWER)

anxious to assist at the family repast, I devoted special attention to watching the mothers eat. As a rule, the prey is consumed out of sight, in the burrow; but sometimes also a meal is taken on the threshold, in the open air. Besides, it is easy to rear the Lycosa and her family in a wire-gauze cage, with a layer of earth wherein the captive will never dream of sinking a well, such work being out of season. Everything then happens in the open.

Well, while the mother munches, chews, expresses the juices and swallows, the youngsters do not budge from their camping-ground on her back. Not one quits its place nor gives a sign of wishing to slip down and join in the meal. Nor does the mother extend an invitation to them to come and recruit themselves, nor put any broken victuals aside for them. She feeds and the others look on, or rather remain indifferent to what is happening. Their perfect quiet during the Lycosa's feast points to the posession of a stomach that knows no cravings.

Then with what are they sustained, during their seven months' upbringing on the mother's back? One conceives a notion of exudations supplied by the bearer's body, in which case the young would feed on their mother, after the manner of parasitic vermin, and gradually drain her strength.

We must abandon this notion. Never are they seen to put their mouths to the skin that should be a sort of teat to them. On the other hand, the Lycosa, far from being exhausted and shrivelling, keeps perfectly well and plump. She has the same pot-belly when she finishes rearing her young as when she began. She has not lost weight: far from it; on the contrary, she has put on flesh: she has gained the wherewithal to beget a new family next summer, one as numerous as to-day's.

Once more, with what do the little ones keep up their strength? We do not like to suggest reserves supplied by the egg as rectifying the beastie's expenditure of vital force, especially when we consider that those reserves, themselves so close to nothing, must be economized in view of the silk, a material of the highest importance, of which a plentiful use will be made presently. There must be other powers at play in the tiny animal's machinery.

Total abstinence from food could be understood, if it were accompanied by inertia: immobility is not life. But the young

Lycosae, although usually quiet on their mother's back, are at all times ready for exercise and for agile swarming. When they fall from the maternal perambulator, they briskly pick themselves up, briskly scramble up a leg and make their way to the top. It is a splendidly nimble and spirited performance. Besides, once seated, they have to keep a firm balance in the mass; they have to stretch and stiffen their little limbs in order to hang on to their neighbours. As a matter of fact, there is no absolute rest for them. Now physiology teaches us that not a fibre works without some expenditure of energy. The animal, which can be likened, in no small measure, to our industrial machines, demands, on the one hand, the renovation of its organism, which wears out with movement, and, on the other, the maintenance of the heat transformed into action. We can compare it with the locomotive-engine. As the iron horse performs its work, it gradually wears out its pistons, its rods, its wheels, its boiler-tubes, all of which have to be made good from time to time. The founder and the smith repair it, supply it, so to speak, with 'plastic food,' the food that becomes embodied with the whole and forms part of it. But, though it have just come from the engine-shop, it is still inert. To acquire the power of movement, it must receive from the stoker a supply of 'energy-producing food;' in other words, he lights a few shovelfuls of coal in its inside. This heat will produce mechanical work.

Even so with the beast. As nothing is made from nothing, the egg supplies first the materials of the new-born animal; then the plastic food, the smith of living creatures, increases the body, up to a certain limit, and renews it as it wears away. The stoker works at the same time, without stopping. Fuel, the source of energy, makes but a short stay in the system, where it is consumed and furnishes heat, whence movement is derived. Life is a fire-box. Warmed by its food, the animal machine moves, walks, runs, jumps, swims, flies, sets its locomotory apparatus going in a thousand manners.

To return to the young Lycosae, they grow no larger until the period of their emancipation. I find them at the age of seven months the same as when I saw them at their birth. The egg supplied the materials necessary for their tiny frames; and, as the loss of waste substance is, for the moment, excessively small, or even *nil*, additional plastic food is not needed so long as the beastie does not grow. In this respect, the prolonged abstinence presents no difficulty. But there remains the question of energy-producing food, which is indispensable, for the little Lycosa moves, when necessary, and very actively at that. To what shall we attribute the heat expended upon action, when the animal takes absolutely no nourishment?

An idea suggests itself. We say to ourselves that, without being life, a machine is something more than matter, for man has added a little of his mind to it. Now the iron beast, consuming its ration of coal, is really browsing the ancient foliage of arborescent ferns in which solar energy has accumulated.

Beasts of flesh and blood act no otherwise. Whether they mutually devour one another or levy tribute on the plant, they invariably quicken themselves with the stimulant of the sun's heat, a heat stored in grass, fruit, seed and those which feed on such. The sun, the soul of the universe, is the supreme dispenser of energy.

Instead of being served up through the intermediary of food and passing through the ignominious circuit of gastric chemistry, could not this solar energy penetrate the animal directly and charge it with activity, even as the battery charges an accumulator with power? Why not live on sun, seeing that, after all, we find naught but sun in the fruits which we consume?

Chemical science, that bold revolutionary, promises to provide us with synthetic food-stuffs. The laboratory and the factory will take the place of the farm. Why should not physical science step in as well? It would leave the preparation of plastic food to the chemist's retorts; it would reserve for itself that of energy-producing food, which, reduced to its exact terms, ceases to be matter. With the aid of some ingenious apparatus, it would pump into us our daily ration of solar energy, to be later expended in movement, whereby the machine would be kept going without the often painful assistance of the stomach and its adjuncts. What a delightful world, where one would lunch off a ray of sunshine!

Is it a dream, or the anticipation of a remote reality? The problem is one of the most important that science can set us. Let us first hear the evidence of the young Lycosae regarding its possibilities.

For seven months, without any material nourishment, they expend strength in moving. To wind up the mechanism of their muscles, they recruit themselves direct with heat and light. During the time when she was dragging the bag of eggs behind her, the mother, at the best moments of the day, came and held up her pill to the sun. With her two hind-legs, she lifted it out of the ground, into the full light; slowly she turned it and returned it, so that every side might receive its share of the vivifying rays. Well, this bath of life, which awakened the germs, is now prolonged to keep the tender babes active.

Daily, if the sky be clear, the Lycosa, carrying her young, comes up from the burrow, leans on the kerb and spends long hours basking in the sun. Here, on their mother's back, the youngsters stretch their limbs delightedly, saturate themselves with heat, take in reserves of motor power, absorb energy.

They are motionless; but, if I only blow upon them, they stampede as nimbly as though a hurricane were passing. Hurriedly, they disperse; hurriedly, they reassemble: a proof that, without material nourishment, the little animal machine is always at full pressure, ready to work. When the shade comes, mother and sons go down again, surfeited with solar emanations. The feast of energy at the Sun Tavern is finished for the day. It is repeated in the same way daily, if the weather be mild, until the hour of emancipation comes, followed by the first mouthfuls of solid food.

CHAPTER VI

THE NARBONNE LYCOSA: THE CLIMBING-INSTINCT

THE month of March comes to an end; and the departure of the youngsters begins, in glorious weather, during the hottest hours of the morning. Laden with her swarming burden, the mother Lycosa is outside her burrow, squatting on the parapet at the entrance. She lets them do as they please; as though indifferent to what is happening, she exhibits neither encouragement nor regret. Whoso will goes; whoso will remains behind.

First these, then those, according as they feel themselves duly soaked with sunshine, the little ones leave the mother in batches, run about for a moment on the ground and then quickly reach the trellis-work of the cage, which they climb with surprising alacrity. They pass through the meshes, they clamber right to the top of the citadel. All, with not one exception, make for the heights, instead of roaming on the ground, as might reasonably be expected from the eminently earthly habits of the Lycosae; all ascend the dome, a strange procedure whereof I do not yet guess the object.

I receive a hint from the upright ring that finishes the top of the cage. The youngsters hurry to it. It represents the porch of their gymnasium. They hang out threads across the opening; they stretch others from the ring to the nearest points of the trellis-work. On these foot-bridges, they perform slack-rope exercises amid endless comings and goings. The tiny legs open out from time to time and straddle as though to reach the most distant points. I begin to realize that they are acrobats aiming at loftier heights than those of the dome.

I top the trellis with a branch that doubles the attainable height. The bustling crowd hastily scrambles up it, reaches the tip of the topmost twigs and thence sends out threads that attach themselves to every surrounding object. These form so many suspension-bridges; and my beasties nimbly run along them, incessantly passing to and fro. One would say that they wished to climb higher still. I will endeavour to satisfy their desires.

I take a nine-foot reed, with tiny branches spreading right up to the top, and place it above the cage. The little Lycosae clamber to the very summit. Here, longer threads are produced from the rope-yard and are now left to float, anon converted into bridges by the mere contact of the free end with the neighbouring supports. The rope-dancers embark upon them and form garlands which the least breath of air swings daintily. The thread is invisible when it does not come between the eyes and the sun; and the whole suggests rows of Gnats dancing an aerial ballet.

Then, suddenly, teased by the air-currents, the delicate mooring breaks and flies through space. Behold the emigrants off and away, clinging to their thread. If the wind be favourable, they can land at great distances. Their departure is thus continued for a week or two, in bands more or less numerous, according to the temperature and the brightness of the day. If the sky be overcast, none dreams of leaving. The travellers need the kisses of the sun, which give energy and vigour.

At last, the whole family has disappeared, carried afar by its flying-ropes. The mother remains alone. The loss of her offspring hardly seems to distress her. She retains her usual colour and plumpness, which is a sign that the maternal exertions have not been too much for her.

I also notice an increased fervour in the chase. While burdened with her family, she was remarkably abstemious, accepting only with great reserve the game placed at her disposal. The coldness of the season may have militated against copious refections; perhaps also the weight of the little ones hampered her movements and made her more discreet in attacking the prey.

To-day, cheered by the fine weather and able to move freely, she hurries up from her lair each time I set a tit-bit to her liking buzzing at the entrance to her burrow; she comes and takes from my fingers the savoury Locust, the portly Anoxia[1]; and this performance is

---

[1] A species of Beetle.—*Translator's Note*.

THE EMIGRANTS, OFF AND AWAY, CLINGING TO THEIR THREAD

repeated daily, whenever I have the leisure to devote to it. After a frugal winter, the time has come for plentiful repasts.

This appetite tells us that the animal is not at the point of death; one does not feast in this way with a played-out stomach. My boarders are entering in full vigour upon their fourth year. In the winter, in the fields, I used to find large mothers, carting their young, and others not much more than half their size. The whole series, therefore, represented three generations. And now, in my earthenware pans, after the departure of the family, the old matrons still carry on and continue as strong as ever. Every outward appearance tells us that, after becoming great-grandmothers, they still keep themselves fit for propagating their species.

The facts correspond with these anticipations. When September returns, my captives are dragging a bag as bulky as that of last year. For a long time, even when the eggs of the others have been hatched for some weeks past, the mothers come daily to the threshold of the burrow and hold out their wallets for incubation by the sun. Their perseverance is not rewarded: nothing issues from the satin purse; nothing stirs within. Why? Because, in the prison of my cages, the eggs have had no father. Tired of waiting and at last recognizing the barrenness of their produce, they push the bag of eggs outside the burrow and trouble about it no more. At the return of spring, by which time the family, if developed according to rule, would have been emancipated, they die. The mighty Spider of the waste-lands, therefore, attains to an even more patriarchal age than her neighbour the Sacred Beetle[1]: she lives for five years at the very least.

Let us leave the mothers to their business and return to the youngsters. It is not without a certain surprise that we see the little Lycosae, at the first moment of their emancipation, hasten to ascend the heights. Destined to live on the ground, amidst the short grass, and afterwards to settle in the permanent abode, a pit, they start by being enthusiastic acrobats. Before descending to the low levels, their normal dwelling-place, they affect lofty altitudes.

To rise higher and ever higher is their first need. I have not, it seems, exhausted the limit of their climbing-instinct even with a

---

[1] Cf. Insect Life, by J. H. Fabre, translated by the author of Mademoiselle Mori: Chaps. I. and II.; The Life and Love of the Insect, by J. Henri Fabre, translated by Alexander Teixeira de Mattos: Chaps. I. to IV.—*Translator's Note*.

nine-foot pole, suitably furnished with branches to facilitate the escalade. Those who have eagerly reached the very top wave their legs, fumble in space as though for yet higher stalks. It behoves us to begin again and under better conditions.

Although the Narbonne Lycosa, with her temporary yearning for the heights, is more interesting than other Spiders, by reason of the fact that her usual habitation is underground, she is not so striking at swarming-time, because the youngsters, instead of all migrating at once, leave the mother at different periods and in small batches. The sight will be a finer one with the common Garden or Cross Spider, the Diadem Epeira (*Epeira diadema*, LIN.), decorated with three white crosses on her back.

She lays her eggs in November and dies with the first cold snap. She is denied the Lycosa's longevity. She leaves the natal wallet early one spring and never sees the following spring. This wallet, which contains the eggs, has none of the ingenious structure which we admired in the Banded and in the Silky Epeira. No longer do we see a graceful balloon-shape nor yet a paraboloid with a starry base; no longer a tough, waterproof satin stuff; no longer a swan's-down resembling a fleecy, russet cloud; no longer an inner keg in which the eggs are packed. The art of stout fabrics and of walls within walls is unknown here.

The work of the Cross Spider is a pill of white silk, wrought into a yielding felt, through which the new-born Spiders will easily work their way, without the aid of the mother, long since dead, and without having to rely upon its bursting at the given hour. It is about the size of a damson.

We can judge the method of manufacture from the structure. Like the Lycosa, whom we saw, in Chapter III., at work in one of my earthenware pans, the Cross Spider, on the support supplied by a few threads stretched between the nearest objects, begins by making a shallow saucer of sufficient thickness to dispense with subsequent corrections. The process is easily guessed. The tip of the abdomen goes up and down, down and up with an even beat, while the worker shifts her place a little. Each time, the spinnerets add a bit of thread to the carpet already made.

When the requisite thickness is obtained, the mother empties her ovaries, in one continuous flow, into the centre of the bowl.

Glued together by their inherent moisture, the eggs, of a handsome orange-yellow, form a ball-shaped heap. The work of the spinnerets is resumed. The ball of germs is covered with a silk cap, fashioned in the same way as the saucer. The two halves of the work are so well joined that the whole constitutes an unbroken sphere.

The Banded Epeira and the Silky Epeira, those experts in the manufacture of rainproof textures, lay their eggs high up, on brushwood and bramble, without shelter of any kind. The thick material of the wallets is enough to protect the eggs from the inclemencies of the winter, especially from damp. The Diadem Epeira, or Cross Spider, needs a cranny for hers, which is contained in a non-waterproof felt. In a heap of stones, well exposed to the sun, she will choose a large slab to serve as a roof. She lodges her pill underneath it, in the company of the hibernating Snail.

More often still, she prefers the thick tangle of some dwarf shrub, standing eight or nine inches high and retaining its leaves in winter. In the absence of anything better, a tuft of grass answers the purpose. Whatever the hiding-place, the bag of eggs is always near the ground, tucked away as well as may be, amid the surrounding twigs.

Save in the case of the roof supplied by a large stone, we see that the site selected hardly satisfies proper hygienic needs. The Epeira seems to realize this fact. By way of an additional protection, even under a stone, she never fails to make a thatched roof for her eggs. She builds them a covering with bits of fine, dry grass, joined together with a little silk. The abode of the eggs becomes a straw wigwam.

Good luck procures me two Cross Spiders' nests, on the edge of one of the paths in the enclosure, among some tufts of ground-cypress, or lavender-cotton. This is just what I wanted for my plans. The find is all the more valuable as the period of the exodus is near at hand.

I prepare two lengths of bamboo, standing about fifteen feet high and clustered with little twigs from top to bottom. I plant one of them straight up in the tuft, beside the first nest. I clear the surrounding ground, because the bushy vegetation might easily, thanks to threads carried by the wind, divert the emigrants from the road which I have laid out for them. The other bamboo I set up in the middle of the yard, all by itself, some few steps from any outstanding object. The second nest is removed as it is, shrub and all, and placed at the bottom of the tall, ragged distaff.

The events expected are not long in coming. In the first fortnight in May, a little earlier in one case, a little later in the other, the two families, each presented with a bamboo climbing-pole, leave their respective wallets. There is nothing remarkable about the mode of egress. The precincts to be crossed consist of a very slack net-work, through which the outcomers wriggle: weak little orange-yellow beasties, with a triangular black patch upon their sterns. One morning is long enough for the whole family to make its appearance.

By degrees, the emancipated youngsters climb the nearest twigs, clamber to the top, and spread a few threads. Soon, they gather in a compact, ball-shaped cluster, the size of a walnut. They remain motionless. With their heads plunged into the heap and their sterns projecting, they doze gently, mellowing under the kisses of the sun. Rich in the possession of a thread in their belly as their sole inheritance, they prepare to disperse over the wide world.

Let us create a disturbance among the globular group by stirring it with a straw. All wake up at once. The cluster softly dilates and spreads, as though set in motion by some centrifugal force; it becomes a transparent orb wherein thousands and thousands of tiny legs quiver and shake, while threads are extended along the way to be followed. The whole work resolves itself into a delicate veil which swallows up the scattered family. We then see an exquisite nebula against whose opalescent tapestry the tiny animals gleam like twinkling orange stars.

This straggling state, though it last for hours, is but temporary. If the air grow cooler, if rain threaten, the spherical group reforms at once. This is a protective measure. On the morning after a shower, I find the families on either bamboo in as good condition as on the day before. The silk veil and the pill formation have sheltered them well enough from the downpour. Even so do Sheep, when caught in a storm in the pastures, gather close, huddle together and make a common rampart of their backs.

The assembly into a ball-shaped mass is also the rule in calm, bright weather, after the morning's exertions. In the afternoon, the climbers collect at a higher point, where they weave a wide, conical tent, with the end of a shoot for its top, and, gathered into a compact group, spend the night there. Next day, when the heat returns, the ascent is resumed in long files, following the shrouds which a few pioneers have rigged and which those who come after elaborate with their own work.

Collected nightly into a globular troop and sheltered under a fresh tent, for three or four days, each morning, before the sun grows too hot, my little emigrants thus raise themselves, stage by stage, on both bamboos, until they reach the sun-unit, at fifteen feet above the ground. The climb comes to an end for lack of foothold.

Under normal conditions, the ascent would be shorter. The young Spiders have at their disposal the bushes, the brushwood, providing supports on every side for the threads wafted hither and thither by the eddying air-currents. With these rope-bridges flung across space, the dispersal presents no difficulties. Each emigrant leaves at his own good time and travels as suits him best.

My devices have changed these conditions somewhat. My two bristling poles stand at a distance from the surrounding shrubs, especially the one which I planted in the middle of the yard. Bridges are out of the question, for the threads flung into the air are not long enough. And so the acrobats, eager to get away, keep on climbing, never come down again, are impelled to seek in a higher position what they have failed to find in a lower. The top of my two bamboos probably fails to represent the limit of what my keen climbers are capable of achieving.

We shall see, in a moment, the object of this climbing-propensity, which is a sufficiently remarkable instinct in the Garden Spiders, who have as their domain the low-growing brushwood wherein their nets are spread; it becomes a still more remarkable instinct in the Lycosa, who, except at the moment when she leaves her mother's back, never quits the ground and yet, in the early hours of her life, shows herself as ardent a wooer of high places as the young Garden Spiders.

Let us consider the Lycosa in particular. In her, at the moment of the exodus, a sudden instinct arises, to disappear, as promptly and for ever, a few hours later. This is the climbing-instinct, which is unknown to the adult and soon forgotten by the emancipated youngling, doomed to wander homeless, for many a long day, upon the ground. Neither of them dreams of climbing to the top of a grass-stalk. The full-grown Spider hunts trapper-fashion, ambushed in her tower; the young one hunts afoot through the scrubby grass. In both cases there is no web and therefore no need for lofty contact-points. They are not allowed to quit the ground and climb the heights.

Yet here we have the young Lycosa, wishing to leave the maternal abode and to travel far afield by the easiest and swiftest methods, suddenly becoming an enthusiastic climber. Impetuously she scales the wire trellis of the cage where she was born; hurriedly she clambers to the top of the tall mast which I have prepared for her. In the same way, she would make for the summit of the bushes in her waste-land.

We catch a glimpse of her object. From on high, finding a wide space beneath her, she sends a thread floating. It is caught by the wind and carries her hanging to it. We have our aeroplanes; she too possesses her flying-machine. Once the journey is accomplished, naught remains of this ingenious business. The climbing-instinct conies suddenly, at the hour of need, and no less suddenly vanishes.

CHAPTER VII

THE SPIDERS' EXODUS

SEEDS, when ripened in the fruit, are disseminated, that is to say, scattered on the surface of the ground, to sprout in spots as yet unoccupied and fill the expanses that realize favourable conditions.

Amid the wayside rubbish grows one of the gourd family, *Ecbalium elaterium*, commonly called the squirting cucumber, whose fruit—a rough and extremely bitter little cucumber—is the size of a date. When ripe, the fleshy core resolves into a liquid in which float the seeds. Compressed by the elastic rind of the fruit, this liquid bears upon the base of the footstalk, which is gradually forced out, yields like a stopper, breaks off and leaves an orifice through which a stream of seeds and fluid pulp is suddenly ejected. If, with a novice hand, under a scorching sun, you shake the plant laden with yellow fruit, you are bound to be somewhat startled when you hear a noise among the leaves and receive the cucumber's grapeshot in your face.

The fruit of the garden balsam, when ripe, splits, at the least touch, into five fleshy valves, which curl up and shoot their seeds to a distance. The botanical name of *Impatiens* given to the balsam alludes to this sudden dehiscence of the capsules, which cannot endure contact without bursting.

In the damp and shady places of the woods there exists a plant of the same family which, for similar reasons, bears the even more expressive name of *Impatiens noli-me-tangere*, or touch-me-not.

The capsule of the pansy expands into three valves, each scooped out like a boat and laden in the middle with two rows of seeds. When these valves dry, the edges shrivel, press upon the grains and eject them.

Light seeds, especially those of the order of Compositae, have aeronautic apparatus—tufts, plumes, fly-wheels—which keep them up in the air and enable them to take distant voyages. In this way, at the least breath, the seeds of the dandelion, surmounted by a tuft of feathers, fly from their dry receptacle and waft gently in the air.

Next to the tuft, the wing is the most satisfactory contrivance for dissemination by wind. Thanks to their membranous edge, which gives them the appearance of thin scales, the seeds of the yellow wall-flower reach high cornices of buildings, clefts of inaccessible rocks, crannies in old walls, and sprout in the remnant of mould bequeathed by the mosses that were there before them.

The samaras, or keys, of the elm, formed of a broad, light fan with the seed cased in its centre; those of the maple, joined in pairs and resembling the unfurled wings of a bird; those of the ash, carved like the blade of an oar, perform the most distant journeys when driven before the storm.

Like the plant, the insect also sometimes possesses travelling-apparatus, means of dissemination that allow large families to disperse quickly over the country, so that each member may have his place in the sun without injuring his neighbour; and these apparatus, these methods vie in ingenuity with the elm's samara, the dandelion-plume and the catapult of the squirting cucumber.

Let us consider, in particular, the Epeirae, those magnificent Spiders who, to catch their prey, stretch, between one bush and the next, great vertical sheets of meshes, resembling those of the fowler. The most remarkable in my district is the Banded Epeira (*Epeira fasciata*, WALCK.), so prettily belted with yellow, black and silvery white. Her nest, a marvel of gracefulness, is a satin bag, shaped like a tiny pear. Its neck ends in a concave mouthpiece closed with a lid, also of satin. Brown ribbons, in fanciful meridian waves, adorn the object from pole to pole.

Open the nest. We have seen, in an earlier chapter[1], what we find there; let us retell the story. Under the outer wrapper, which is as stout as our woven stuffs and, moreover, perfectly waterproof, is a russet eiderdown of exquisite delicacy, a silky fluff resembling driven smoke. Nowhere does mother-love prepare a softer bed.

---

[1]Chapter II.—*Translator's Note*.

In the middle of this downy mass hangs a fine, silk, thimble-shaped purse, closed with a movable lid. This contains the eggs, of a pretty orange-yellow and about five hundred in number.

All things considered, is not this charming edifice an animal fruit, a germ-casket, a capsule to be compared with that of the plants? Only, the Epeira's wallet, instead of seeds, holds eggs. The difference is more apparent than real, for egg and grain are one.

How will this living fruit, ripening in the heat beloved of the Cicadae, manage to burst? How, above all, will dissemination take place? They are there in their hundreds. They must separate, go far away, isolate themselves in a spot where there is not too much fear of competition among neighbours. How will they set to work to achieve this distant exodus, weaklings that they are, taking such very tiny steps?

I receive the first answer from another and much earlier Epeira, whose family I find, at the beginning of May, on a yucca in the enclosure. The plant blossomed last year. The branching flower-stem, some three feet high, still stands erect, though withered. On the green leaves, shaped like a sword-blade, swarm two newly-hatched families. The wee beasties are a dull yellow, with a triangular black patch upon their stern. Later on, three white crosses, ornamenting the back, will tell me that my find corresponds with the Cross or Diadem Spider (*Epeira diadema*, WALCK.).

When the sun reaches this part of the enclosure, one of the two groups falls into a great state of flutter. Nimble acrobats that they are, the little Spiders scramble up, one after the other, and reach the top of the stem. Here, marches and countermarches, tumult and confusion reign, for there is a slight breeze which throws the troop into disorder. I see no connected manoeuvres. From the top of the stalk they set out at every moment, one by one; they dart off suddenly; they fly away, so to speak. It is as though they had the wings of a Gnat.

Forthwith they disappear from view. Nothing that my eyes can see explains this strange flight; for precise observation is impossible amid the disturbing influences out of doors. What is wanted is a peaceful atmosphere and the quiet of my study.

I gather the family in a large box, which I close at once, and instal it in the animals' laboratory, on a small table, two steps from the open window. Apprised by what I have just seen of their propensity to resort to the heights, I give my subjects a bundle of twigs, eighteen inches tall, as a climbing-pole. The whole band hurriedly clambers up and reaches the top. In a few moments there is not one lacking in the group on high. The future will tell us the reason of this assemblage on the projecting tips of the twigs.

The little Spiders are now spinning here and there at random: they go up, go down, come up again. Thus is woven a light veil of divergent threads, a many-cornered web with the end of the branch for its summit and the edge of the table for its base, some eighteen inches wide. This veil is the drill-ground, the work-yard where the preparations for departure are made.

Here hasten the humble little creatures, running indefatigably to and fro. When the sun shines upon them, they become gleaming specks and form upon the milky background of the veil a sort of constellation, a reflex of those remote points in the sky where the telescope shows us endless galaxies of stars. The immeasurably small and the immeasurably large are alike in appearance. It is all a matter of distance.

But the living nebula is not composed of fixed stars; on the contrary, its specks are in continual movement. The young Spiders never cease shifting their position on the web. Many let themselves drop, hanging by a length of thread, which the faller's weight draws from the spinnerets. Then quickly they climb up again by the same thread, which they wind gradually into a skein and lengthen by successive falls. Others confine themselves to running about the web and also give me the impression of working at a bundle of ropes.

The thread, as a matter of fact, does not flow from the spinneret; it is drawn thence with a certain effort. It is a case of extraction, not emission. To obtain her slender cord, the Spider has to move about and haul, either by falling or by walking, even as the rope-maker steps backwards when working his hemp. The activity now displayed on the drill-ground is a preparation for the approaching dispersal. The travellers are packing up.

Soon we see a few Spiders trotting briskly between the table and the open window. They are running in mid-air. But on what? If the light fall favourably, I manage to see, at moments, behind the tiny animal, a thread resembling a ray of light, which appears for an instant, gleams and disappears. Behind, therefore, there is a mooring, only just perceptible, if you look very carefully; but, in front, towards the window, there is nothing to be seen at all.

In vain I examine above, below, at the side; in vain I vary the direction of the eye: I can distinguish no support for the little creature to walk upon. One would think that the beastie were paddling in space. It suggests the idea of a small bird, tied by the leg with a thread and making a flying rush forwards.

But, in this case, appearances are deceptive: flight is impossible; the Spider must necessarily have a bridge whereby to cross the intervening space. This bridge, which I cannot see, I can at least destroy. I cleave the air with a ruler in front of the Spider making for the window. That is quite enough: the tiny animal at once ceases to go forward and falls. The invisible foot-plank is broken. My son, young Paul, who is helping me, is astounded at this wave of the magic wand, for not even he, with his fresh, young eyes, is able to see a support ahead for the Spiderling to move along.

In the rear, on the other hand, a thread is visible. The difference is easily explained. Every Spider, as she goes, at the same time spins a safety-cord which will guard the rope-walker against the risk of an always possible fall. In the rear, therefore, the thread is of double thickness and can be seen, whereas, in front, it is still single and hardly perceptible to the eye.

Obviously, this invisible foot-bridge is not flung out by the animal: it is carried and unrolled by a gust of air. The Epeira, supplied with this line, lets it float freely; and the wind, however softly blowing, bears it along and unwinds it. Even so is the smoke from the bowl of a pipe whirled up in the air.

This floating thread has but to touch any object in the neighbourhood and it will remain fixed to it. The suspension-bridge is thrown; and the Spider can set out. The South-American Indians are said to cross the abysses of the Cordilleras in travelling-cradles made of twisted creepers; the little Spider passes through space on the invisible and the imponderable.

But to carry the end of the floating thread elsewhither a draught is needed. At this moment, the draught exists between the door of my study and the window, both of which are open. It is so slight that I do not feel its; I only know of it by the smoke from my pipe, curling softly in that direction. Cold air enters from without through the door; warm air escapes from the room through the window. This is the drought that carries the threads with it and enables the Spiders to embark upon their journey.

I get rid of it by closing both apertures and I break off any communication by passing my ruler between the window and the table. Henceforth, in the motionless atmosphere, there are no departures. The current of air is missing, the skeins are not unwound and migration becomes impossible.

It is soon resumed, but in a direction whereof I never dreamt. The hot sun is beating on a certain part of the floor. At this spot, which is warmer than the rest, a column of lighter, ascending air is generated. If this column catch the threads, my Spiders ought to rise to the ceiling of the room.

The curious ascent does, in fact, take place. Unfortunately, my troop, which has been greatly reduced by the number of departures through the window, does not lend itself to prolonged experiment. We must begin again.

The next morning, on the same yucca, I gather the second family, as numerous as the first. Yesterday's preparations are repeated. My legion of Spiders first weaves a divergent framework between the top of the brushwood placed at the emigrants' disposal and the edge of the table. Five or six hundred wee beasties swarm all over this work-yard.

While this little world is busily fussing, making its arrangements for departure, I make my own. Every aperture in the room is closed, so as to obtain as calm an atmosphere as possible. A small chafing-dish is lit at the foot of the table. My hands cannot feel the heat of it at the level of the web whereon my Spiders are weaving. This is the very modest fire which, with its column of rising air, shall unwind the threads and carry them on high.

Let us first enquire the direction and strength of the current. Dandelion-plumes, made lighter by the removal of their seeds, serve as my guides. Released above the chafing-dish, on the level of the table, they float slowly upwards and, for the most part, reach the ceiling. The emigrants' lines should rise in the same way and even better.

The thing is done: with the aid of nothing that is visible to the three of us looking on, a Spider makes her ascent. She ambles with her eight legs through the air; she mounts, gently swaying. The others, in ever-increasing numbers, follow, sometimes by different roads, sometimes by the same road. Any one who did not possess the secret would stand amazed at this magic ascent without a ladder. In a few minutes, most of them are up, clinging to the ceiling.

Not all of them reach it. I see some who, on attaining a certain height, cease to go up and even lose ground, although moving their legs forward with all the nimbleness of which they are capable. The more they struggle upwards, the faster they come down. This drifting, which neutralizes the distance covered and even converts it into a retrogression, is easily explained.

The thread has not reached the platform; it floats, it is fixed only at the lower end. As long as it is of a fair length, it is able, although moving, to bear the minute animal's weight. But, as the Spider climbs, the float becomes shorter in proportion; and the time comes when a balance is struck between the ascensional force of the thread and the weight carried. Then the beastie remains stationary, although continuing to climb.

Presently, the weight becomes too much for the shorter and shorter float; and the Spider slips down, in spite of her persistent, forward striving. She is at last brought back to the branch by the falling threads. Here, the ascent is soon renewed, either on a fresh thread, if the supply of silk be not yet exhausted, or on a strange thread, the work, of those who have gone before.

As a rule, the ceiling is reached. It is twelve feet high. The little Spider is able, therefore, as the first product of her spinning-mill, before taking any refreshment, to obtain a line fully twelve feet in length. And all this, the rope-maker and her rope, was contained in the egg, a particle of no size at all. To what a degree of fineness can the silky matter be wrought wherewith the young Spider is provided! Our manufacturers are able to turn out platinum-wire that can only be seen when it is made red-hot. With much simpler means, the Spiderling draws from her wire-mill threads so delicate that, even the brilliant light of the sun does not always enable us to discern them.

We must not let all the climbers be stranded on the ceiling, an inhospitable region where most of them will doubtless perish, being unable to produce a second thread before they have had a meal. I open the window. A current of lukewarm air, coming from the chafing-dish, escapes through the top. Dandelion-plumes, taking that direction, tell me so. The wafting threads cannot fail to be carried by this flow of air and to lengthen out in the open, where a light breeze is blowing.

I take a pair of sharp scissors and, without shaking the threads, cut a few that are just visible at the base, where they are thickened with an added strand. The result of this operation is marvellous. Hanging to the flying-rope, which is borne on the wind outside, the Spider passes through the window, suddenly flies off and disappears. An easy way of travelling, if the conveyance possessed a rudder that allowed the passenger to land where he pleases! But the little things are at the mercy of the winds: where will they alight? Hundreds, thousands of yards away, perhaps. Let us wish them a prosperous journey.

The problem of dissemination is now solved. What would happen if matters, instead of being brought about by my wiles, took place in the open fields? The answer is obvious. The young Spiders, born acrobats and rope-walkers, climb to the top of a branch so as to find sufficient space below them to unfurl their apparatus. Here, each draws from her rope-factory a thread which she abandons to the eddies of the air. Gently raised by the currents that ascend from the ground warmed by the sun, this thread wafts upwards, floats, undulates, makes for its point of contact. At last, it breaks and vanishes in the distance, carrying the spinstress hanging to it.

The Epeira with the three white crosses, the Spider who has supplied us with these first data concerning the process of dissemination, is endowed with a moderate maternal industry. As a receptacle for the eggs, she weaves a mere pill of silk. Her work is modest indeed beside the Banded Epeira's balloons. I looked to these to supply me with fuller documents. I had laid up a store by rearing some mothers during the autumn. So that nothing of importance might escape me, I divided my stock of balloons, most of which were woven before my eyes, into two sections. One half remained in my study, under a wire-gauze cover, with, small bunches of brushwood as supports; the other half were experiencing the vicissitudes of open-air life on the rosemaries in the enclosure.

These preparations, which promised so well, did not provide me with the sight which I expected, namely, a magnificent exodus, worthy of the tabernacle occupied. However, a few results, not devoid of interest, are to be noted. Let us state them briefly.

The hatching takes place as March approaches. When this time comes, let us open the Banded Epeira's nest with the scissors. We shall find that some of the youngsters have already left the central chamber and scattered over the surrounding eiderdown, while the rest of the laying still consists of a compact mass of orange eggs. The appearance of the younglings is not simultaneous; it takes place with intermissions and may last a couple of weeks.

Nothing as yet suggests the future, richly-striped livery. The abdomen is white and, as it were, floury in the front half; in the other half it is a blackish-brown. The rest of the body is pale-yellow, except in front, where the eyes form a black edging. When left alone, the little ones remain motionless in the soft, russet swan's-down; if disturbed, they shuffle lazily where they are, or even walk about in a hesitating and unsteady fashion. One can see that they have to ripen before venturing outside.

Maturity is achieved in the exquisite floss that surrounds the natal chamber and fills out the balloon. This is the waiting-room in which the body hardens. All dive into it as and when they emerge from the central keg. They will not leave it until four months later, when the midsummer heats have come.

Their number is considerable. A patient and careful census gives me nearly six hundred. And all this comes out of a purse no larger than a pea. By what miracle is there room for such a family? How do those thousands of legs manage to grow without straining themselves?

The egg-bag, as we learnt in Chapter II., is a short cylinder rounded at the bottom. It is formed of compact white satin, an insuperable barrier. It opens into a round orifice wherein is bedded a lid of the same material, through which the feeble beasties would be incapable of passing. It is not a porous felt, but a fabric as tough as that of the sack. Then by what mechanism is the delivery effected?

Observe that the disk of the lid doubles back into a short fold, which edges into the orifice of the bag. In the same way, the lid of a saucepan fits the mouth by means of a projecting rim, with this difference, that the rim is not attached to the saucepan, whereas, in the Epeira's work, it is soldered to the bag or nest. Well, at the time of the hatching, this disk becomes unstuck, lifts and allows the new-born Spiders to pass through.

If the rim were movable and simply inserted, if, moreover, the birth of all the family took place at the same time, we might think that the door is forced open by the living wave of inmates, who would set their backs to it with a common effort. We should find an approximate image in the case of the saucepan, whose lid is raised by the boiling of its contents. But the fabric of the cover is one with the fabric of the bag, the two are closely welded; besides, the hatching is effected in small batches, incapable of the least exertion. There must, therefore, be a spontaneous bursting, or dehiscence, independent of the assistance of the youngsters and similar to that of the seed-pods of plants.

When fully ripened, the dry fruit of the snap-dragon opens three windows; that of the pimpernel splits into two rounded halves, something like those of the outer case of a fob-watch; the fruit of the carnation partly unseals its valves and opens at the top into a star-shaped hatch. Each seed-casket has its own system of locks, which are made to work smoothly by the mere kiss of the sun.

Well, that other dry fruit, the Banded Epeira's germ-box, likewise possesses its bursting-gear. As long as the eggs remain unhatched, the door, solidly fixed in its frame, holds good; as soon as the little ones swarm and want to get out, it opens of itself.

Come June and July, beloved of the Cicadae, no less beloved of the young Spiders who are anxious to be off. It were difficult indeed for them to work their way through the thick shell of the balloon. For the second time, a spontaneous dehiscence seems called for. Where will it be effected?

The idea occurs off-hand that it will take place along the edges of the top cover. Remember the details given in an earlier chapter. The neck of the balloon ends in a wide crater, which is closed by a ceiling

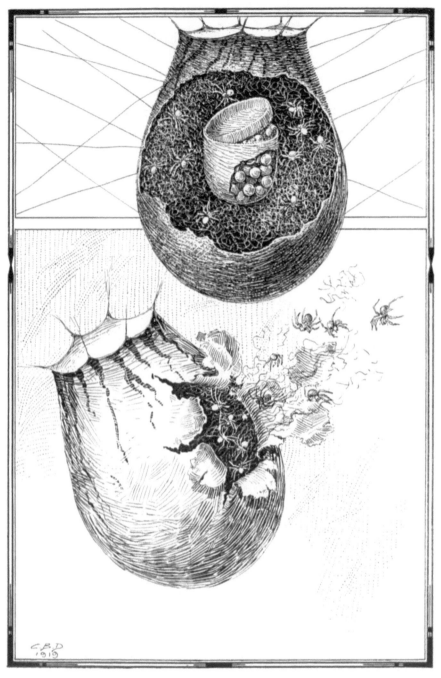

THIS DISC BECOMES UNSTUCK, LIFTS AND ALLOWS THE NEW-BORN SPIDER TO PASS THROUGH (UPPER)

THE BALLOONS OF THE BANDED EPEIRA... BURST UNDER THE RAYS OF A TORRID SUN (LOWER)

dug out cup-wise. The material is as stout in this part as in any other; but, as the lid was the finishing touch to the work, we expect to find an incomplete soldering, which would allow it to be unfastened.

The method of construction deceives us: the ceiling is immovable; at no season can my forceps manage to extract it, without destroying the building from top to bottom. The dehiscence takes place elsewhere, at some point on the sides. Nothing informs us, nothing suggests to us that it will occur at one place rather than another.

Moreover, to tell the truth, it is not a dehiscence prepared by means of some dainty piece of mechanism; it is a very irregular tear. Somewhat sharply, under the fierce heat of the sun, the satin bursts like the rind of an over-ripe pomegranate. Judging by the result, we think of the expansion of the air inside, which, heated by the sun, causes this rupture. The signs of pressure from within are manifest: the tatters of the torn fabric are turned outwards; also, a wisp of the russet eiderdown that fills the wallet invariably straggles through the breach. In the midst of the protruding floss, the Spiderlings, expelled from their home by the explosion, are in frantic commotion.

The balloons of the Banded Epeira are bombs which, to free their contents, burst under the rays of a torrid sun. To break they need the fiery heat-waves of the dog-days. When kept in the moderate atmosphere of my study, most of them do not open and the emergence of the young does not take place, unless I myself I have a hand in the business; a few others open with a round hole, a hole so neat that it might have been made with a punch. This aperture is the work of the prisoners, who, relieving one another in turns, have, with a patient tooth, bitten through the stuff of the jar at some point or other.

When exposed to the full force of the sun, however, on the rosemaries in the enclosure, the balloons burst and shoot forth a ruddy flood of floss and tiny animals. That is how things occur in the free sun-bath of the fields. Unsheltered, among the bushes, the wallet of the Banded Epeira, when the July heat arrives, splits under the effort of the inner air. The delivery is effected by an explosion of the dwelling.

A very small part of the family are expelled with the flow of tawny floss; the vast majority remain in the bag, which is ripped open, but still bulges with eiderdown. Now that the breach is made, any one can go out who pleases, in his own good time, without hurrying. Besides, a solemn action has to be performed before the emigration. The animal must cast its skin; and the moult is an event that does not fall on the same date for all. The evacuation of the place, therefore, lasts several days. It is effected in small squads, as the slough is flung aside.

Those who sally forth climb up the neighbouring twigs and there, in the full heat of the sun, proceed with the work of dissemination. The method is the same as that which we saw in the case of the Cross Spider. The spinnerets abandon to the breeze a thread that floats, breaks and flies away, carrying the rope-maker with it. The number of starters on any one morning is so small as to rob the spectacle of the greater part of its interest. The scene lacks animation because of the absence of a crowd.

To my intense disappointment, the Silky Epeira does not either indulge in a tumultuous and dashing exodus. Let me remind you of her handiwork, the handsomest of the maternal wallets, next to the Banded Epeira's. It is an obtuse conoid, closed with a star-shaped disk. It is made of a stouter and especially a thicker material than the Banded Epeira's balloon, for which reason a spontaneous rupture becomes more necessary than ever.

This rupture is effected at the sides of the bag, not far from the edge of the lid. Like the ripping of the balloon, it requires the rough aid of the heat of July. Its mechanism also seems to work by the expansion of the heated air, for we again see a partial emission of the silky floss that fills the pouch.

The exit of the family is performed in a single group and, this time, before the moult, perhaps for lack of the space necessary for the delicate casting of the skin. The conical bag falls far short of the balloon in size; those packed within would sprain their legs in extracting them from their sheaths. The family, therefore, emerges in a body and settles on a sprig hard by.

This is a temporary camping-ground, where, spinning in unison, the youngsters soon weave an open-work tent, the abode of a week, or thereabouts. The moult is effected in this lounge of intersecting threads. The sloughed skins form a heap at the bottom of the dwelling; on the trapezes above, the flaylings take exercise and gain strength and vigour. Finally, when maturity is attained, they set out, now these, now those, little by little and always cautiously. There are no audacious flights on the thready airship; the journey is accomplished by modest stages.

Hanging to her thread, the Spider lets herself drop straight down, to a depth of nine or ten inches. A breath of air sets her swinging like a pendulum, sometimes drives her against a neighbouring branch. This is a step towards the dispersal. At the point reached, there is a fresh fall, followed by a fresh pendulous swing that lands her a little farther afield. Thus, in short tacks, for the thread is never very long, does the Spiderling go about, seeing the country, until she comes to a place that suits her. Should the wind blow at all hard, the voyage is cut short: the cable of the pendulum breaks and the beastie is carried for some distance on its cord.

To sum up, although, on the whole, the tactics of the exodus remain much the same, the two spinstresses of my region best-versed in the art of weaving mothers' wallets failed to come up to my expectations. I went to the trouble of rearing them, with disappointing results. Where shall I find again the wonderful spectacle which the Cross Spider offered me by chance? I shall find it—in an even more striking fashion—among humbler Spiders, whom I had neglected to observe.

CHAPTER VIII

THE CRAB SPIDER

THE Spider that showed me the exodus in all its magnificence is known officially as *Thomisus onustus*, WALCK. Though the name suggest nothing to the reader's mind, it has the advantage, at any rate, of hurting neither the throat nor the ear, as is too often the case with scientific nomenclature, which sounds more like sneezing than articulate speech. Since it is the rule to dignify plants and animals with a Latin label, let us at least respect the euphony of the classics and refrain from harsh splutters which spit out a name instead of pronouncing it.

What will posterity do in face of the rising tide of a barbarous vocabulary which, under the pretence of progress, stifles real knowledge? It will relegate the whole business to the quagmire of oblivion. But what will never disappear is the popular name, which sounds well, is picturesque and conveys some sort of information. Such is the term Crab Spider, applied by the ancients to the group to which the Thomisus belongs, a pretty accurate term, for, in this case, there is an evident analogy between the Spider and the Crustacean.

Like the Crab, the Thomisus walks sideways; she also has forelegs stronger than her hind-legs. The only thing wanting to complete the resemblance is the front pair of stone gauntlets, raised in the attitude of self-defence.

The Spider with the Crab-like figure does not know how to manufacture nets for catching game. Without springs or snares, she lies in ambush, among the flowers, and awaits the arrival of the quarry, which she kills by administering a scientific stab in the neck. The Thomisus, in particular, the subject of this chapter, is passionately addicted to the pursuit of the Domestic Bee. I have described the contests between the victim and her executioner, at greater length, elsewhere.

The Bee appears, seeking no quarrel, intent upon plunder. She tests the flowers with her tongue; she selects a spot that will yield a good return. Soon she is wrapped up in her harvesting. While she is filling her baskets and distending her crop, the Thomisus, that bandit lurking under cover of the flowers, issues from her hiding-place, creeps round behind the bustling insect, steals up close and, with a sudden rush, nabs her in the nape of the neck. In vain, the Bee protests and darts her sting at random; the assailant does not let go.

Besides, the bite in the neck is paralysing, because the cervical nerve-centres are affected. The poor thing's legs stiffen; and all is over in a second. The murderess now sucks the victim's blood at her ease and, when she has done, scornfully flings the drained corpse aside. She hides herself once more, ready to bleed a second gleaner should the occasion offer.

This slaughter of the Bee engaged in the hallowed delights of labour has always revolted me. Why should there be workers to feed idlers, why sweated to keep sweaters in luxury? Why should so many admirable lives be sacrificed to the greater prosperity of brigandage? These hateful discords amid the general harmony perplex the thinker, all the more as we shall see the cruel vampire become a model of devotion where her family is concerned.

The ogre loved his children; he ate the children of others. Under the tyranny of the stomach, we are all of us, beasts and men alike, ogres. The dignity of labour, the joy of life, maternal affection, the terrors of death: all these do not count, in others; the main point is that morsel the be tender and savoury.

According to the etymology of her name—θωμιγξ, a cord—the Thomisus should be like the ancient lictor, who bound the sufferer to the stake. The comparison is not inappropriate as regards many Spiders who tie their prey with a thread to subdue it and consume it at their ease; but it just happens that the Thomisus is at variance with her label. She does not fasten her Bee, who, dying suddenly of a bite in the neck, offers no resistance to her consumer. Carried away by his recollection of the regular tactics, our Spider's godfather

THE BEE APPEARS SEEKING NO QUARREL (UPPER)
SHE GENTLY LET'S HERSELF DIE, HUGGING HER NEST (LOWER)

overlooked the exception; he did not know of the perfidious mode of attack which renders the use of a bow-string superfluous.

Nor is the second name of *onustus*—loaded, burdened, freighted—any too happily chosen. The fact that the Bee-huntress carries a heavy paunch is no reason to refer to this as a distinctive characteristic. Nearly all Spiders have a voluminous belly, a silk-warehouse where, in some cases, the rigging of the net, in others, the swan's-down of the nest is manufactured. The Thomisus, a first-class nest-builder, does like the rest: she hoards in her abdomen, but without undue display of obesity, the wherewithal to house her family snugly.

Can the expression *onustus* refer simply to her slow and sidelong walk? The explanation appeals to me, without satisfying me fully. Except in the case of a sudden alarm, every Spider maintains a sober gait and a wary pace. When all is said, the scientific term is composed of a misconception and a worthless epithet. How difficult it is to name animals rationally! Let us be indulgent to the nomenclator: the dictionary is becoming exhausted and the constant flood that requires cataloguing mounts incessantly, wearing out our combinations of syllables.

As the technical name tells the reader nothing, how shall he be informed? I see but one means, which is to invite him to the May festivals, in the waste-lands of the South. The murderess of the Bees is of a chilly constitution; in our parts, she hardly ever moves away from the olive-districts. Her favourite shrub is the white-leaved rock-rose (*Cistus albidus*), with the large, pink, crumpled, ephemeral blooms that last but a morning and are replaced, next day, by fresh flowers, which have blossomed in the cool dawn. This glorious efflorescence goes on for five or six weeks.

Here, the Bees plunder enthusiastically, fussing and bustling in the spacious whorl of the stamens, which beflour them with yellow. Their persecutrix knows of this affluence. She posts herself in her watch-house, under the rosy screen of a petal. Cast your eyes over the flower, more or less everywhere. If you see a Bee lying lifeless, with legs and tongue out-stretched, draw nearer: the Thomisus will be there, nine times out of ten. The thug has struck her blow; she is draining the blood of the departed.

After all, this cutter of Bees' throats is a pretty, a very pretty creature, despite her unwieldy paunch fashioned like a squat pyramid and embossed on the base, on either side, with a pimple shaped like a camel's hump. The skin, more pleasing to the eye than any satin, is milk-white in some, in others lemon-yellow. There are fine ladies among them who adorn their legs with a number of pink bracelets and their back with carmine arabesques. A narrow pale-green ribbon sometimes edges the right and left of the breast. It is not so rich as the costume of the Banded Epeira, but much more elegant because of its soberness, its daintiness and the artful blending of its hues. Novice fingers, which shrink from touching any other Spider, allow themselves to be enticed by these attractions; they do not fear to handle the beauteous Thomisus, so gentle in appearance.

Well, what can this gem among Spiders do? In the first place, she makes a nest worthy of its architect. With twigs and horse-hair and bits of wool, the Goldfinch, the Chaffinch and other masters of the builder's art construct an aerial bower in the fork of the branches. Herself a lover of high places, the Thomisus selects as the site of her nest one of the upper twigs of the rock-rose, her regular hunting-ground, a twig withered by the heat and possessing a few dead leaves, which curl into a little cottage. This is where she settles with a view to her eggs.

Ascending and descending with a gentle swing in more or less every direction, the living shuttle, swollen with silk, weaves a bag whose outer casing becomes one with the dry leaves around. The work, which is partly visible and partly hidden by its supports, is a pure dead-white. Its shape, moulded in the angular interval between the bent leaves, is that of a cone and reminds us, on a smaller scale, of the nest of the Silky Epeira.

When the eggs are laid, the mouth of the receptacle is hermetically closed with a lid of the same white silk. Lastly, a few threads, stretched like a thin curtain, form a canopy above the nest and, with the curved tips of the leaves, frame a sort of alcove wherein the mother takes up her abode.

It is more than a place of rest after the fatigues of her confinement: it is a guard-room, an inspection-post where the mother remains sprawling until the youngsters' exodus. Greatly emaciated by the laying of her eggs and by her expenditure of silk, she lives only for the protection of her nest.

Should some vagrant pass near by, she hurries from her watch-tower, lifts a limb and puts the intruder to flight. If I tease her with a straw, she parries with big gestures, like those of a prize-fighter. She uses her fists against my weapon. When I propose to dislodge her in view of certain experiments, I find some difficulty in doing so. She clings to the silken floor, she frustrates my attacks, which I am bound to moderate lest I should injure her. She is no sooner attracted outside than she stubbornly returns to her post. She declines to leave her treasure.

Even so does the Narbonne Lycosa struggle when we try to take away her pill. Each displays the same pluck and the same devotion; and also the same denseness in distinguishing her property from that of others. The Lycosa accepts without hesitation any strange pill which she is, given in exchange for her own; she confuses alien produce with the produce of her ovaries and her silk-factory. Those hallowed words, maternal love, were out of place here: it is an impetuous, an almost mechanical impulse, wherein real affection plays no part whatever. The beautiful Spider of the rock-roses is no more generously endowed. When moved from her nest to another of the same kind, she settles upon it and never stirs from it, even though the different arrangement of the leafy fence be such as to warn her that she is not really at home. Provided that she have satin under her feet, she does not notice her mistake; she watches over another's nest with the same vigilance which she might show in watching over her own.

The Lycosa surpasses her in maternal blindness. She fastens to her spinnerets and dangles, by way of a bag of eggs, a ball of cork polished with my file, a paper pellet, a little ball of thread. In order to discover if the Thomisus is capable of a similar error, I gathered some broken pieces of silk-worm's cocoon into a closed cone, turning the fragments so as to bring the smoother and more delicate inner surface outside. My attempt was unsuccessful. When removed from her home and placed on the artificial wallet, the mother Thomisus obstinately refused to settle there. Can she be more clear-sighted than the Lycosa? Perhaps so. Let us not be too extravagant with our praise, however; the imitation of the bag was a very clumsy one.

The work of laying is finished by the end of May, after which, lying flat on the ceiling of her nest, the mother never leaves her guard-room, either by night or day. Seeing her look so thin and wrinkled, I imagine that I can please her by bringing her a provision of Bees, as I was wont to do. I have misjudged her needs. The Bee, hitherto her favourite dish, tempts her no longer. In vain does the prey buzz close by, an easy capture within the cage: the watcher does not shift from her post, takes no notice of the windfall. She lives exclusively upon maternal devotion, a commendable but unsubstantial fare. And so I see her pining away from day to day, becoming more and more wrinkled. What is the withered thing waiting for, before expiring? She is waiting for her children to emerge; the dying creature is still of use to them.

When the Banded Epeira's little ones issue from their balloon, they have long been orphans. There is none to come to their assistance; and they have not the strength to free themselves unaided. The balloon has to split automatically and to scatter the youngsters and their flossy mattress all mixed up together. The Thomisus' wallet, sheathed in leaves over the greater part of its surface, never bursts; nor does the lid rise, so carefully is it sealed down. Nevertheless, after the delivery of the brood, we see, at the edge of the lid, a small, gaping hole, an exit-window. Who contrived this window, which was not there at first?

The fabric is too thick and tough to have yielded to the twitches of the feeble little prisoners. It was the mother, therefore, who, feeling her offspring shuffle impatiently under the silken ceiling, herself made a hole in the bag. She persists in living for five or six weeks, despite her shattered health, so as to give a last helping hand and open the door for her family. After performing this duty, she gently lets herself die, hugging her nest and turning into a shrivelled relic.

When July comes, the little ones emerge. In view of their acrobatic habits, I have placed a bundle of slender twigs at the top of the cage in which they were born. All of them pass through the wire gauze and form a group on the summit of the brushwood, where they swiftly weave a spacious lounge of criss-cross threads.

Here they remain, pretty quietly, for a day or two; then foot-bridges begin to be flung from one object to the next. This is the opportune moment.

I put the bunch laden with beasties on a small table, in the shade, before the open window. Soon, the exodus commences, but slowly and unsteadily. There are hesitations, retrogressions, perpendicular falls at the end of a thread, ascents that bring the hanging Spider up again. In short much ado for a poor result.

As matters continue to drag, it occurs to me, at eleven o'clock, to take the bundle of brushwood swarming with the little Spiders, all eager to be off, and place it on the window-sill, in the glare of the sun. After a few minutes of heat and light, the scene assumes a very different aspect. The emigrants run to the top of the twigs, bustle about actively. It becomes a bewildering rope-yard, where thousands of legs are drawing the hemp from the spinnerets. I do not see the ropes manufactured and sent floating at the mercy of the air; but I guess their presence.

Three or four Spiders start at a time, each going her own way in directions independent of her neighbours'. All are moving upwards, all are climbing some support, as can be perceived by the nimble motion of their legs. Moreover, the road is visible behind the climber, it is of double thickness, thanks to an added thread. Then, at a certain height, individual movement ceases. The tiny animal soars in space and shines, lit up by the sun. Softly it sways, then suddenly takes flight.

What has happened? There is a slight breeze outside. The floating cable has snapped and the creature has gone off, borne on its parachute. I see it drifting away, showing, like a spot of light, against the dark foliage of the near cypresses, some forty feet distant. It rises higher, it crosses over the cypress-screen, it disappears. Others follow, some higher, some lower, hither and thither.

But the throng has finished its preparations; the hour has come to disperse in swarms. We now see, from the crest of the brushwood, a continuous spray of starters, who shoot up like microscopic projectiles and mount in a spreading cluster. In the end, it is like the bouquet at the finish of a pyrotechnic display, the sheaf of rockets fired simultaneously. The comparison is correct down to the dazzling light itself. Flaming in the sun like so many gleaming points, the little Spiders are the sparks of that living firework. What a glorious send-off! What an entrance into the world! Clutching its aeronautic thread, the minute creature mounts in an apotheosis.

Sooner or later, nearer or farther, the fall comes. To live, we have to descend, often very low, alas! The Crested Lark crumbles the mule-droppings in the road and thus picks up his food, the oaten grain which he would never find by soaring in the sky, his throat swollen with song. We have to descend; the stomach's inexorable claims demand it. The Spiderling, therefore, touches land. Gravity, tempered by the parachute, is kind to her.

The rest of her story escapes me. What infinitely tiny Midges does she capture before possessing the strength to stab her Bee? What are the methods, what the wiles of atom contending with atom? I know not. We shall find her again in spring, grown quite large and crouching among the flowers whence the Bee takes toll.

CHAPTER IX

THE GARDEN SPIDERS: BUILDING THE WEB

THE fowling-snare is one of man's ingenious villainies. With lines, pegs and poles, two large, earth-coloured nets are stretched upon the ground, one to the right, the other to the left of a bare surface. A long cord, pulled, at the right moment, by the fowler, who hides in a brushwood hut, works them and brings them together suddenly, like a pair of shutters.

Divided between the two nets are the cages of the decoy-birds—Linnets and Chaffinches, Greenfinches and Yellowhammers, Buntings and Ortolans—sharp-eared creatures which, on perceiving the distant passage of a flock of their own kind, forthwith utter a short calling note. One of them, the *Sambé*, an irresistible tempter, hops about and flaps his wings in apparent freedom. A bit of twine fastens him to his convict's stake. When, worn with fatigue and driven desperate by his vain attempts to get away, the sufferer lies down flat and refuses to do his duty, the fowler is able to stimulate him without stirring from his hut. A long string sets in motion a little lever working on a pivot. Raised from the ground by this diabolical contrivance, the bird flies, falls down and flies up again at each jerk of the cord.

The fowler waits, in the mild sunlight of the autumn morning. Suddenly, great excitement in the cages. The Chaffinches chirp their rallying-cry:

'Pinck! Pinck!'

There is something happening in the sky. The *Sambé*, quick! They are coming, the simpletons; they swoop down upon the treacherous floor. With a rapid movement, the man in ambush pulls his string. The nets close and the whole flock is caught.

Man has wild beast's blood in his veins. The fowler hastens to the slaughter. With his thumb, he stifles the beating of the captives' hearts, staves in their skulls. The little birds, so many piteous heads of game, will go to market, strung in dozens on a wire passed through their nostrils.

For scoundrelly ingenuity the Epeira's net can bear comparison with the fowler's; it even surpasses it when, on patient study, the main features of its supreme perfection stand revealed. What refinement of art for a mess of Flies! Nowhere, in the whole animal kingdom, has the need to eat inspired a more cunning industry. If the reader will meditate upon the description that follows, he will certainly share my admiration.

First of all, we must witness the making of the net; we must see it constructed and see it again and again, for the plan of such a complex work can only be grasped in fragments. To-day, observation will give us one detail; to-morrow, it will give us a second, suggesting fresh points of view; as our visits multiply, a new fact is each time added to the sum total of the acquired data, confirming those which come before or directing our thoughts along unsuspected paths.

The snow-ball rolling over the carpet of white grows enormous, however scanty each fresh layer be. Even so with truth in observational science: it is built up of trifles patiently gathered together. And, while the collecting of these trifles means that the student of Spider industry must not be chary of his time, at least it involves no distant and speculative research. The smallest garden contains Epeirae, all accomplished weavers.

In my enclosure, which I have stocked carefully with the most famous breeds, I have six different species under observation, all of a useful size, all first-class spinners. Their names are the Banded Epeira (*Epeira fasciata*, WALCK.), the Silky Epeira (*E. sericea*, WALCK.), the Angular Epeira (*E. angulata*, WALCK.), the Pale-tinted Epeira (*E. pallida*, OLIV.), the Diadem Epeira, or Cross Spider (*E. diadema*, CLERK.), and the Crater Epeira (*E. cratera*, WALCK.).

I am able, at the proper hours, all through the fine season, to question them, to watch them at work, now this one, anon that, according to the chances of the day. What I did not see very plainly yesterday I can see the next day, under better conditions, and on any of the following days, until the phenomenon under observation is revealed in all clearness.

Let us go every evening, step by step, from one border of tall rosemaries to the next. Should things move too slowly, we will sit down at the foot of the shrubs, opposite the rope-yard, where the light falls favourably, and watch with unwearying attention. Each trip will be good for a fact that fills some gap in the ideas already gathered. To appoint one's self, in this way, an inspector of Spiders' webs, for many years in succession and for long seasons, means joining a not overcrowded profession, I admit. Heaven knows, it does not enable one to put money by! No matter: the meditative mind returns from that school fully satisfied.

To describe the separate progress of the work in the case of each of the six Epeirae mentioned would be a useless repetition: all six employ the same methods and weave similar webs, save for certain details that shall be set forth later. I will, therefore, sum up in the aggregate the particulars supplied by one or other of them.

My subjects, in the first instance, are young and boast but a slight corporation, very far removed from what it will be in the late autumn. The belly, the wallet containing the rope-works, hardly exceeds a peppercorn in bulk. This slenderness on the part of the spinstresses must not prejudice us against their work: there is no parity between their skill and their years. The adult Spiders, with their disgraceful paunches, can do no better.

Moreover, the beginners have one very precious advantage for the observer: they work by day, work even in the sun, whereas the old ones weave only at night, at unseasonable hours. The first show us the secrets of their looms without much difficulty; the others conceal them from us. Work starts in July, a couple of hours before sunset.

The spinstresses of my enclosure then leave their daytime hiding-places, select their posts and begin to spin, one here, another there. There are many of them; we can choose where we please. Let us stop in front of this one, whom we surprise in the act of laying the foundations of the structure. Without any appreciable order, she runs about the rosemary-hedge, from the tip of one branch to another within the limits of some eighteen inches. Gradually, she puts a thread in position, drawing it from her wire-mill with the combs attached to her hind-legs. This preparatory work presents no appearance of a concerted plan. The Spider comes and goes impetuously, as though at random; she goes up, comes down, goes up again, dives down again and each time strengthens the points of contact with intricate moorings distributed here and there. The result is a scanty and disordered scaffolding.

Is disordered the word? Perhaps not. The Epeira's eye, more experienced in matters of this sort than mine, has recognized the general lie of the land; and the rope-fabric has been erected accordingly: it is very inaccurate in my opinion, but very suitable for the Spider's designs. What is it that she really wants? A solid frame to contain the network of the web. The shapeless structure which she has just built fulfils the desired conditions: it marks out a flat, free and perpendicular area. This is all that is necessary.

The whole work, for that matter, is now soon completed; it is done all over again, each evening, from top to bottom, for the incidents of the chase destroy it in a night. The net is as yet too delicate to resist the desperate struggles of the captured prey. On the other hand, the adults' net, which is formed of stouter threads, is adapted to last some time; and the Epeira gives it a more carefully-constructed framework, as we shall see elsewhere.

A special thread, the foundation of the real net, is stretched across the area so capriciously circumscribed. It is distinguished from the others by its isolation, its position at a distance from any twig that might interfere with its swaying length. It never fails to have, in the middle, a thick white point, formed of a little silk cushion. This is the beacon that marks the centre of the future edifice, the post that will guide the Epeira and bring order into the wilderness of twists and turns.

The time has come to weave the hunting-snare. The Spider starts from the centre, which bears the white signpost, and, running along the transversal thread, hurriedly reaches the circumference, that is to say, the irregular frame enclosing the free space. Still with the same sudden movement, she rushes from the circumference to the centre; she starts again backwards and forwards, makes for the right, the left, the top, the bottom; she hoists herself up, dives down, climbs up again, runs down and always returns to the central landmark by roads that slant in the most unexpected manner. Each

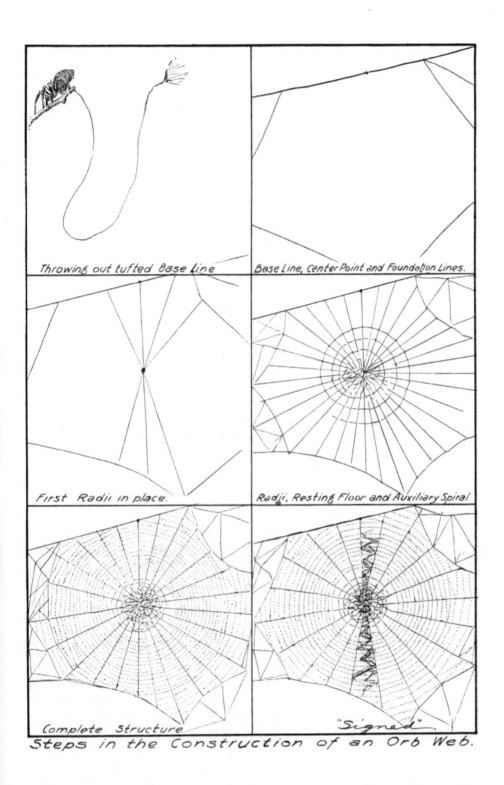

time, a radius or spoke is laid, here, there, or elsewhere, in what looks like mad disorder.

The operation is so erratically conducted that it takes the most unremitting attention to follow it at all. The Spider reaches the margin of the area by one of the spokes already placed. She goes along this margin for some distance from the point at which she landed, fixes her thread to the frame and returns to the centre by the same road which she has just taken.

The thread obtained on the way in a broken line, partly on the radius and partly on the frame, is too long for the exact distance between the circumference and the central point. On returning to this point, the Spider adjusts her thread, stretches it to the correct length, fixes it and collects what remains on the central signpost. In the case of each radius laid, the surplus is treated in the same fashion, so that the signpost continues to increase in size. It was first a speck; it is now a little pellet, or even a small cushion of a certain breadth.

We shall see presently what becomes of this cushion whereon the Spider, that niggardly housewife, lays her saved-up bits of thread; for the moment, we will note that the Epeira works it up with her legs after placing each spoke, teazles it with her claws, mats it into felt with noteworthy diligence. In so doing, she gives the spokes a solid common support, something like the hub of our carriage-wheels.

The eventual regularity of the work suggests that the radii are spun in the same order in which they figure in the web, each following immediately upon its next neighbour. Matters pass in another manner, which at first looks like disorder, but which is really a judicious contrivance. After setting a few spokes in one direction, the Epeira runs across to the other side to draw some in the opposite direction. These sudden changes of course are highly logical; they show us how proficient the Spider is in the mechanics of rope-construction. Were they to succeed one another regularly, the spokes of one group, having nothing as yet to counteract them, would distort the work by their straining, would even destroy it for lack of a stabler support. Before continuing, it is necessary to lay a converse group which will maintain the whole by its resistance. Any combination of forces acting in one direction must be forthwith neutralized by another in the opposite direction. This is what our statics teach us and what the Spider puts into practice; she is a past mistress of the secrets of rope-building, without serving an apprenticeship.

One would think that this interrupted and apparently disordered labour must result in a confused piece of work. Wrong: the rays are equidistant and form a beautifully-regular orb. Their number is a characteristic mark of the different species. The Angular Epeira places 21 in her web, the Banded Epeira 32, the Silky Epeira 42. These numbers are not absolutely fixed; but the variation is very slight.

Now which of us would undertake, off-hand, without much preliminary experiment and without measuring-instruments, to divide a circle into a given quantity of sectors of equal width? The Epeirae, though weighted with a wallet and tottering on threads shaken by the wind, effect the delicate division without stopping to think. They achieve it by a method which seems mad according to our notions of geometry. Out of disorder they evolve order.

We must not, however, give them more than their due. The angles are only approximately equal; they satisfy the demands of the eye, but cannot stand the test of strict measurement. Mathematical precision would be superfluous here. No matter, we are amazed at the result obtained. How does the Epeira come to succeed with her difficult problem, so strangely managed? I am still asking myself the question.

The laying of the radii is finished. The Spider takes her place in the centre, on the little cushion formed of the inaugural signpost and the bits of thread left over. Stationed on this support, she slowly turns round and round. She is engaged on a delicate piece of work. With an extremely thin thread, she describes from spoke to spoke, starting from the centre, a spiral line with very close coils. The central space thus worked attains, in the adults' webs, the dimensions of the palm of one's hand; in the younger Spiders' webs, it is much smaller, but it is never absent. For reasons which I will explain in the course of this study, I shall call it, in future, the 'resting-floor.'

The thread now becomes thicker. The first could hardly be seen; the second is plainly visible. The Spider shifts her position with great slanting strides, turns a few times, moving farther and farther from the centre, fixes her line each time to the spoke which she crosses and at last comes to a stop at the lower edge of the frame. She has described a spiral with coils of rapidly-increasing width. The average distance between the coils, even in the structures of the young Epeirae, is one centimetre[1].

Let us not be misled by the word 'spiral,' which conveys the notion of a curved line. All curves are banished from the Spiders' work; nothing is used but the straight line and its combinations. All that is aimed at is a polygonal line drawn in a curve as geometry understands it. To this polygonal line, a work destined to disappear as the real toils are woven, I will give the name of the 'auxiliary spiral.' Its object is to supply cross-bars, supporting rungs, especially in the outer zone, where the radii are too distant from one another to afford a suitable groundwork. Its object is also to guide the Epeira in the extremely delicate business which she is now about to undertake.

But, before that, one last task becomes essential. The area occupied by the spokes is very irregular, being marked out by the supports of the branch, which are infinitely variable. There are angular niches which, if skirted too closely, would disturb the symmetry of the web about to be constructed. The Epeira needs an exact space wherein gradually to lay her spiral thread. Moreover, she must not leave any gaps through which her prey might find an outlet.

An expert in these matters, the Spider soon knows the corners that have to be filled up. With an alternating movement, first in this direction, then in that, she lays, upon the support of the radii, a thread that forms two acute angles at the lateral boundaries of the faulty part and describes a zigzag line not wholly unlike the ornament known as the fret.

The sharp corners have now been filled with frets on every side; the time has come to work at the essential part, the snaring-web for which all the rest is but a support. Clinging on the one hand to the radii, on the other to the chords of the auxiliary spiral, the Epeira covers the same ground as when laying the spiral, but in the opposite direction: formerly, she moved away from the centre; now

---

[1] .39 inch.—*Translator's Note*.

she moves towards it and with closer and more numerous circles. She starts from the base of the auxiliary spiral, near the frame.

What follows is difficult to observe, for the movements are very quick and spasmodic, consisting of a series of sudden little rushes, sways and bends that bewilder the eye. It needs continuous attention and repeated examination to distinguish the progress of the work however slightly.

The two hind-legs, the weaving implements, keep going constantly. Let us name them according to their position on the work-floor. I call the leg that faces the centre of the coil, when the animal moves, the 'inner leg;' the one outside the coil the 'outer leg.'

The latter draws the thread from the spinneret and passes it to the inner leg, which, with a graceful movement, lays it on the radius crossed. At the same time, the first leg measures the distance; it grips the last coil placed in position and brings within a suitable range that point of the radius whereto the thread is to be fixed. As soon as the radius is touched, the thread sticks to it by its own glue. There are no slow operations, no knots: the fixing is done of itself.

Meanwhile, turning by narrow degrees, the spinstress approaches the auxiliary chords that have just served as her support. When, in the end, these chords become too close, they will have to go; they would impair the symmetry of the work. The Spider, therefore, clutches and holds on to the rungs of a higher row; she picks up, one by one, as she goes along, those which are of no more use to her and gathers them into a fine-spun ball at the contact-point of the next spoke. Hence arises a series of silky atoms marking the course of the disappearing spiral.

The light has to fall favourably for us to perceive these specks, the only remains of the ruined auxiliary thread. One would take them for grains of dust, if the faultless regularity of their distribution did not remind us of the vanished spiral. They continue, still visible, until the final collapse of the net.

And the Spider, without a stop of any kind, turns and turns and turns, drawing nearer to the centre and repeating the operation of fixing her thread at each spoke which she crosses. A good half-hour, an hour even among the full-grown Spiders, is spent on spiral circles, to the number of about fifty for the web of the Silky Epeira and thirty for those of the Banded and the Angular Epeira.

At last, at some distance from the centre, on the borders of what I have called the resting-floor, the Spider abruptly terminates her spiral when the space would still allow of a certain number of turns. We shall see the reason of this sudden stop presently. Next, the Epeira, no matter which, young or old, hurriedly flings herself upon the little central cushion, pulls it out and rolls it into a ball which I expected to see thrown away. But no: her thrifty nature does not permit this waste. She eats the cushion, at first an inaugural landmark, then a heap of bits of thread; she once more melts in the digestive crucible what is no doubt intended to be restored to the silken treasury. It is a tough mouthful, difficult for the stomach to elaborate; still, it is precious and must not be lost. The work finishes with the swallowing. Then and there, the Spider instals herself, head downwards, at her hunting-post in the centre of the web.

The operation which we have just seen gives rise to a reflection. Men are born right-handed. Thanks to a lack of symmetry that has never been explained, our right side is stronger and readier in its movements than our left. The inequality is especially noticeable in the two hands. Our language expresses this supremacy of the favoured side in the terms dexterity, adroitness and address, all of which allude to the right hand.

Is the animal, on its side, right-handed, left-handed, or unbiased? We have had opportunities of showing that the Cricket, the Grasshopper and many others draw their bow, which is on the right wing-case, over the sounding apparatus, which is on the left wing-case. They are right-handed.

When you and I take an unpremeditated turn, we spin round on our right heel. The left side, the weaker, moves on the pivot of the right, the stronger. In the same way, nearly all the Molluscs that have spiral shells roll their coils from left to right. Among the numerous species in both land and water fauna, only a very few are exceptional and turn from right to left.

It would be interesting to try and work out to what extent that part of the zoological kingdom which boasts a two-sided structure is divided into right-handed and left-handed animals. Can dissymmetry, that source of contrasts, be a general rule? Or are there neutrals, endowed with equal powers of skill and energy on both sides? Yes, there are; and the Spider is one of them. She enjoys the very enviable privilege of possessing a left side which is no less capable than the right. She is ambidextrous, as witness the following observations.

When laying her snaring-thread, every Epeira turns in either direction indifferently, as a close watch will prove. Reasons whose secret escapes us determine the direction adopted. Once this or the other course is taken, the spinstress does not change it, even after incidents that sometimes occur to disturb the progress of the work. It may happen that a Gnat gets caught in the part already woven. The Spider thereupon abruptly interrupts her labours, hastens up to the prey, binds it and then returns to where she stopped and continues the spiral in the same order as before.

At the commencement of the work, gyration in one direction being employed as well as gyration in the other, we see that, when making her repeated webs, the same Epeira turns now her right side, now her left to the centre of the coil. Well, as we have said, it is always with the inner hind-leg, the leg nearer the centre, that is to say, in some cases the right and in some cases the left leg, that she places the thread in position, an exceedingly delicate operation calling for the display of exquisite skill, because of the quickness of the action and the need for preserving strictly equal distances. Any one seeing this leg working with such extreme precision, the right leg to-day, the left to-morrow, becomes convinced that the Epeira is highly ambidextrous.

CHAPTER X

THE GARDEN SPIDERS: MY NEIGHBOUR

AGE does not modify the Epeira's talent in any essential feature. As the young worked, so do the old, the richer by a year's experience. There are no masters nor apprentices in their guild; all know their craft from the moment that the first thread is laid. We have learnt something from the novices: let us now look into the matter of their elders and see what additional task the needs of age impose upon them.

July comes and gives me exactly what I wish for. While the new inhabitants are twisting their ropes on the rosemaries in the enclosure, one evening, by the last gleams of twilight, I discover a splendid Spider, with a mighty belly, just outside my door. This one is a matron; she dates back to last year; her majestic corpulence, so exceptional at this season, proclaims the fact. I know her for the Angular Epeira (*Epeira angulata*, WALCK.), clad in grey and girdled with two dark stripes that meet in a point at the back. The base of her abdomen swells into a short nipple on either side.

This neighbour will certainly serve my turn, provided that she do not work too late at night. Things bode well: I catch the buxom one in the act of laying her first threads. At this rate my success need not be won at the expense of sleep. And, in fact, I am able, throughout the month of July and the greater part of August, from eight to ten o'clock in the evening, to watch the construction of the web, which is more or less ruined nightly by the incidents of the chase and built up again, next day, when too seriously dilapidated.

During the two stifling months, when the light fails and a spell of coolness follows upon the furnace-heat of the day, it is easy for me, lantern in hand, to watch my neighbour's various operations. She has taken up her abode, at a convenient height for observation, between a row of cypress-trees and a clump of laurels, near the entrance to an alley haunted by Moths. The spot appears well-chosen, for the Epeira does not change it throughout the season, though she renews her net almost every night.

Punctually as darkness falls, our whole family goes and calls upon her. Big and little, we stand amazed at her wealth of belly and her exuberant somersaults in the maze of quivering ropes; we admire the faultless geometry of the net as it gradually takes shape. All agleam in the lantern-light, the work becomes a fairy orb, which seems woven of moonbeams.

Should I linger, in my anxiety to clear up certain details, the household, which by this time is in bed, waits for my return before going to sleep:

'What has she been doing this evening?' I am asked. 'Has she finished her web? Has she caught a Moth?'

I describe what has happened. To-morrow, they will be in a less hurry to go to bed: they will want to see everything, to the very end. What delightful, simple evenings we have spent looking into the Spider's workshop!

The journal of the Angular Epeira, written up day by day, teaches us, first of all, how she obtains the ropes that form the framework of the building. All day invisible, crouching amid the cypress-leaves, the Spider, at about eight o'clock in the evening, solemnly emerges from her retreat and makes for the top of a branch. In this exalted position, she sits for some time laying her plans with due regard to the locality; she consults the weather, ascertains if the night will be fine. Then, suddenly, with her eight legs wide-spread, she lets herself drop straight down, hanging to the line that issues from her spinnerets. Just as the rope-maker obtains the even output of his hemp by walking backwards, so does the Epeira obtain the discharge of hers by falling. It is extracted by the weight of her body.

The descent, however, has not the brute speed which the force of gravity would give it, if uncontrolled. It is governed by the action of the spinnerets, which contract or expand their pores, or close them entirely, at the faller's pleasure. And so, with gentle moderation she pays out this living plumb-line, of which my lantern clearly shows me the plumb, but not always the line. The great squab seems at such times to be sprawling in space, without the least support.

ALL AGLEAM IN THE LANTERN-LIGHT, THE WORK BECOMES A FAIRY ORB, WHICH SEEMS WOVEN OF MOONBEAMS

She comes to an abrupt stop two inches from the ground; the silk-reel ceases working. The Spider turns round, clutches the line which she has just obtained and climbs up by this road, still spinning. But, this time, as she is no longer assisted by the force of gravity, the thread is extracted in another manner. The two hind-legs, with a quick alternate action, draw it from the wallet and let it go.

On returning to her starting-point, at a height of six feet or more, the Spider is now in possession of a double line, bent into a loop and floating loosely in a current of air. She fixes her end where it suits her and waits until the other end, wafted by the wind, has fastened its loop to the adjacent twigs.

The desired result may be very slow in coming. It does not tire the unfailing patience of the Epeira, but it soon wears out mine. And it has happened to me sometimes to collaborate with the Spider. I pick up the floating loop with a straw and lay it on a branch, at a convenient height. The foot-bridge erected with my assistance is considered satisfactory, just as though the wind had placed it. I count this collaboration among the good actions standing to my credit.

Feeling her thread fixed, the Epeira runs along it repeatedly, from end to end, adding a fibre to it on each journey. Whether I help or not, this forms the 'suspension-cable,' the main piece of the framework. I call it a cable, in spite of its extreme thinness, because of its structure. It looks as though it were single, but, at the two ends, it is seen to divide and spread, tuft-wise, into numerous constituent parts, which are the product of as many crossings. These diverging fibres, with their several contact-points, increase the steadiness of the two extremities.

The suspension-cable is incomparably stronger than the rest of the work and lasts for an indefinite time. The web is generally shattered after the night's hunting and is nearly always rewoven on the following evening. After the removal of the wreckage, it is made all over again, on the same site, cleared of everything except the cable from which the new network is to hang.

The laying of this cable is a somewhat difficult matter, because the success of the enterprise does not depend upon the animal's industry alone. It has to wait until a breeze carries the line to the pier-head in the bushes. Sometimes, a calm prevails; sometimes, the thread catches at an unsuitable point. This involves great expenditure of time, with no certainty of success. And so, when once the suspension-cable is in being, well and solidly placed, the Epeira does not change it, except on critical occasions. Every evening, she passes and repasses over it, strengthening it with fresh threads.

When the Epeira cannot manage a fall of sufficient depth to give her the double line with its loop to be fixed at a distance, she employs another method. She lets herself down and then climbs up again, as we have already seen; but, this time, the thread ends suddenly in a filmy hair-pencil, a tuft, whose parts remain disjoined, just as they come from the spinneret's rose. Then this sort of bushy fox's brush is cut short, as though with a pair of scissors, and the whole thread, when unfurled, doubles its length, which is now enough for the purpose. It is fastened by the end joined to the Spider; the other floats in the air, with its spreading tuft, which easily tangles in the bushes. Even so must the Banded Epeira go to work when she throws her daring suspension-bridge across a stream.

Once the cable is laid, in this way or in that, the Spider is in possession of a base that allows her to approach or withdraw from the leafy piers at will. From the height of the cable, the upper boundary of the projected works, she lets herself slip to a slight depth, varying the points of her fall. She climbs up again by the line produced by her descent. The result of the operation is a double thread which is unwound while the Spider walks along her big footbridge to the contact-branch, where she fixes the free end of her thread more or less low down. In this way, she obtains, to right and left, a few slanting cross-bars, connecting the cable with the branches.

These cross-bars, in their turn, support others in ever-changing directions. When there are enough of them, the Epeira need no longer resort to falls in order to extract her threads; she goes from one cord to the next, always wire-drawing with her hind-legs and placing her produce in position as she goes. This results in a combination of straight lines owning no order, save that they are kept in one, nearly perpendicular plane. They mark a very irregular polygonal area, wherein the web, itself a work of magnificent regularity, shall presently be woven.

It is unnecessary to go over the construction of the masterpiece again; the younger Spiders have taught us enough in this respect. In both cases, we see the same equidistant radii laid, with a central landmark for a guide; the same auxiliary spiral, the scaffolding of temporary rungs, soon doomed to disappear; the same snaring-spiral, with its maze of closely-woven coils. Let us pass on: other details call for our attention.

The laying of the snaring-spiral is an exceedingly delicate operation, because of the regularity of the work. I was bent upon knowing whether, if subjected to the din of unaccustomed sounds, the Spider would hesitate and blunder. Does she work imperturbably? Or does she need undisturbed quiet? As it is, I know that my presence and that of my light hardly trouble her at all. The sudden flashes emitted by my lantern have no power to distract her from her task. She continues to turn in the light even as she turned in the dark, neither faster nor slower. This is a good omen for the experiment which I have in view.

The first Sunday in August is the feast of the patron saint of the village, commemorating the Finding of St. Stephen. This is Tuesday, the third day of the rejoicings. There will be fireworks to-night, at nine o'clock, to conclude the merry-makings. They will take place on the high-road outside my door, at a few steps from the spot where my Spider is working. The spinstress is busy upon her great spiral at the very moment when the village big-wigs arrive with trumpet and drum and small boys carrying torches.

More interested in animal psychology than in pyrotechnical displays, I watch the Epeira's doings, lantern in hand. The hullabaloo of the crowd, the reports of the mortars, the crackle of Roman candles bursting in the sky, the hiss of the rockets, the rain of sparks, the sudden flashes of white, red or blue light: none of this disturbs the worker, who methodically turns and turns again, just as she does in the peace of ordinary evenings.

Once before, the gun which I fired under the plane-trees failed to trouble the concert of the Cicadae; to-day, the dazzling light of the fire-wheels and the splutter of the crackers do not avail to distract the Spider from her weaving. And, after all, what difference would it make to my neighbour if the world fell in! The village could be blown up with dynamite, without her losing her head for such a trifle. She would calmly go on with her web.

Let us return to the Spider manufacturing her net under the usual tranquil conditions. The great spiral has been finished, abruptly, on the confines of the resting-floor. The central cushion, a mat of ends of saved thread, is next pulled up and eaten. But, before indulging in this mouthful, which closes the proceedings, two Spiders, the only two of the order, the Banded and the Silky Epeira, have still to sign their work. A broad, white ribbon is laid, in a thick zigzag, from the centre to the lower edge of the orb. Sometimes, but not always, a second band of the same shape and of lesser length occupies the upper portion, opposite the first.

I like to look upon these odd flourishes as consolidating-gear. To begin with, the young Epeirae never use them. For the moment, heedless of the future and lavish of their silk, they remake their web nightly, even though it be none too much dilapidated and might well serve again. A brand-new snare at sunset is the rule with them. And there is little need for increased solidity when the work has to be done again on the morrow.

On the other hand, in the late autumn, the full-grown Spiders, feeling laying-time at hand, are driven to practise economy, in view of the great expenditure of silk required for the egg-bag. Owing to its large size, the net now becomes a costly work which it were well to use as long as possible, for fear of finding one's reserves exhausted when the time comes for the expensive construction of the nest. For this reason, or for others which escape me, the Banded and the Silky Epeirae think it wise to produce durable work and to strengthen their toils with a cross-ribbon. The other Epeirae, who are put to less expense in the fabrication of their maternal wallet—a mere pill—are unacquainted with the zigzag binder and, like the younger Spiders, reconstruct their web almost nightly.

My fat neighbour, the Angular Epeira, consulted by the light of a lantern, shall tell us how the renewal of the net proceeds. As the twilight fades, she comes down cautiously from her day-dwelling; she leaves the foliage of the cypresses for the suspension-cable of her snare. Here she stands for some time; then, descending to her web, she collects the wreckage in great armfuls. Everything—spiral, spokes and frame—is raked up with her legs. One thing alone is spared and that is the suspension-cable, the sturdy piece of work that has served as a foundation for the previous buildings and will serve for the new after receiving a few strengthening repairs.

The collected ruins form a pill which the Spider consumes with the same greed that she would show in swallowing her prey. Nothing remains. This is the second instance of the Spiders' supreme economy of their silk. We have seen them, after the manufacture of the net, eating the central guide-post, a modest mouthful; we now see them gobbling up the whole web, a meal. Refined and turned into fluid by the stomach, the materials of the old net will serve for other purposes.

As goon as the site is thoroughly cleared, the work of the frame and the net begins on the support of the suspension-cable which was respected. Would it not be simpler to restore the old web, which might serve many times yet, if a few rents were just repaired? One would say so; but does the Spider know how to patch her work, as a thrifty housewife darns her linen? That is the question.

To mend severed meshes, to replace broken threads, to adjust the new to the old, in short, to restore the original order by assembling the wreckage would be a far-reaching feat of prowess, a very fine proof of gleams of intelligence, capable of performing rational calculations. Our menders excel in this class of work. They have as their guide their sense, which measures the holes, cuts the new piece to size and fits it into its proper place. Does the Spider possess the counterpart of this habit of clear thinking?

People declare as much, without, apparently, looking into the matter very closely. They seem able to dispense with the conscientious observer's scruples, when inflating their bladder of theory. They go straight ahead; and that is enough. As for ourselves, less greatly daring, we will first enquire; we will see by experiment if the Spider really knows how to repair her work.

The Angular Epeira, that near neighbour who has already supplied me with so many documents, has just finished her web, at nine o'clock in the evening. It is a splendid night, calm and warm, favourable to the rounds of the Moths. All promises good hunting. At the moment when, after completing the great spiral, the Epeira is about to eat the central cushion and settle down upon her resting-floor, I cut the web in two, diagonally, with a pair of sharp scissors. The sagging of the spokes, deprived of their counter-agents, produces an empty space, wide enough for three fingers to pass through.

The Spider retreats to her cable and looks on without being greatly frightened. When I have done, she quietly returns. She takes her stand on one of the halves, at the spot which was the centre of the original orb; but, as her legs find no footing on one side, she soon realizes that the snare is defective. Thereupon, two threads are stretched across the breach, two threads, no more; the legs that lacked a foothold spread across them; and henceforth the Epeira moves no more, devoting her attention to the incidents of the chase.

When I saw those two threads laid, joining the edges of the rent, I began to hope that I was to witness a mending-process:

'The Spider,' said I to myself, 'will increase the number of those cross-threads from end to end of the breach; and, though the added piece may not match the rest of the work, at least it will fill the gap and the continuous sheet will be of the same use practically as the regular web.'

The reality did not answer to my expectation. The spinstress made no further endeavour all night. She hunted with her riven net, for what it was worth; for I found the web next morning in the same condition wherein I had left it on the night before. There had been no mending of any kind.

The two threads stretched across the breach even must not be taken for an attempt at repairing. Finding no foothold for her legs on one side, the Spider went to look into the state of things and, in so doing, crossed the rent. In going and returning, she left a thread, as is the custom with all the Epeirae when walking. It was not a deliberate mending, but the mere result of an uneasy change of place.

Perhaps the subject of my experiment thought it unnecessary to go to fresh trouble and expense, for the web can serve quite well as it is, after my scissor-cut: the two halves together represent the original snaring-surface. All that the Spider, seated in a central position, need do is to find the requisite support for her spread legs. The two threads stretched from side to side of the cleft supply her with this, or nearly. My mischief did not go far enough. Let us devise something better.

Next day, the web is renewed, after the old one has been swallowed. When the work is done and the Epeira seated motionless at her central post, I take a straw and, wielding it dexterously, so as to respect the resting-floor and the spokes, I pull and root up the spiral, which dangles in tatters. With its snaring-threads ruined, the net is useless; no passing Moth would allow herself to be caught. Now what does the Epeira do in the face of this disaster? Nothing at all. Motionless on her resting-floor, which I have left intact, she awaits the capture of the game; she awaits it all night in vain on her impotent web. In the morning, I find the snare as I left it. Necessity, the mother of invention, has not prompted the Spider to make a slight repair in her ruined toils.

Possibly this is asking too much of her resources. The silk-glands may be exhausted after the laying of the great spiral; and to repeat the same expenditure immediately is out of the question. I want a case wherein there could be no appeal to any such exhaustion. I obtain it, thanks to my assiduity.

While I am watching the rolling of the spiral, a head of game rushes fun tilt into the unfinished snare. The Epeira interrupts her work, hurries to the giddy-pate, swathes him and takes her fill of him where he lies. During the struggle, a section of the web has torn under the weaver's very eyes. A great gap endangers the satisfactory working of the net. What will the spider do in the presence of this grievous rent?

Now or never is the time to repair the broken threads: the accident has happened this very moment, between the animal's legs; it is certainly known and, moreover, the rope-works are in full swing. This time there is no question of the exhaustion of the silk-warehouse.

Well, under these conditions, so favourable to darning, the Epeira does no mending at all. She flings aside her prey, after taking a few sips at it, and resumes her spiral at the point where she interrupted it to attack the Moth. The torn part remains as it is. The machine-shuttle in our looms does not revert to the spoiled fabric; even so with the Spider working at her web.

And this is no case of distraction, of individual carelessness; all the large spinstresses suffer from a similar incapacity for patching. The Banded Epeira and the Silky Epeira are noteworthy in this respect. The Angular Epeira remakes her web nearly every evening; the other two reconstruct theirs only very seldom and use them even when extremely dilapidated. They go on hunting with shapeless rags. Before they bring themselves to weave a new web, the old one has to be ruined beyond recognition. Well, I have often noted the state of one of these ruins and, the next morning, I have found it as it was, or even more dilapidated. Never any repairs; never; never. I am sorry, because of the reputation which our hard-pressed theorists have given her, but the Spider is absolutely unable to mend her work. In spite of her thoughtful appearance, the Epeira is incapable of the modicum of reflexion required to insert a piece into an accidental gap.

Other Spiders are unacquainted with wide-meshed nets and weave satins wherein the threads, crossing at random, form a continuous substance. Among this number is the House Spider (*Tegenaria domestica*, LIN.). In the corners of our rooms, she stretches wide webs fixed by angular extensions. The best-protected nook at one side contains the owner's secret apartment. It is a silk tube, a gallery with a conical opening, whence the Spider, sheltered from the eye, watches events. The rest of the fabric, which exceeds our finest muslins in delicacy, is not, properly speaking, a hunting-implement: it is a platform whereon the Spider, attending to the affairs of her estate, goes her rounds, especially at night. The real trap consists of a confusion of lines stretched above the web.

The snare, constructed according to other rules than in the case of the Epeirae, also works differently. Here are no viscous threads, but plain toils, rendered invisible by the very number. If a Gnat rush into the perfidious entanglement, he is caught at once; and the more he struggles the more firmly is he bound. The snareling falls on the sheet-web. *Tegenaria* hastens up and bites him in the neck.

Having said this, let us experiment a little. In the web of the House Spider, I make a round hole, two fingers wide. The hole remains yawning all day long; but next morning it is invariably closed. An extremely thin gauze covers the breach, the dark appearance of which contrasts with the dense whiteness of the surrounding fabric. The gauze is so delicate that, to make sure of its presence, I use a straw rather than my eyes. The movement of the web, when this part is touched, proves the presence of an obstacle.

Here, the matter would appear obvious. The House Spider has mended her work during the night; she has put a patch in the torn stuff, a talent unknown to the Garden Spiders. It would be greatly to her credit, if a mere attentive study did not lead to another conclusion.

The web of the House Spider is, as we were saying, a platform for watching and exploring; it is also a sheet into which the insects caught in the overhead rigging fall. This surface, a domain subject to unlimited shocks, is never strong enough, especially as it is exposed to the additional burden of little bits of plaster loosened from the wall. The owner is constantly working at it; she adds a new layer nightly.

Every time that she issues from her tubular retreat or returns to it, she fixes the thread that hangs behind her upon the road covered. As evidence of this work, we have the direction of the surface-lines, all of which, whether straight or winding, according to the fancies that guide the Spider's path, converge upon the entrance of the tube. Each step taken, beyond a doubt, adds a filament to the web.

We have here the story of the Processionary of the Pine[1], whose habits I have related elsewhere. When the caterpillars leave the silk pouch, to go and browse at night, and also when they enter it again, they never fail to spin a little on the surface of their nest. Each expedition adds to the thickness of the wall.

When moving this way or that upon the purse which I have split from top to bottom with my scissors, the Processionaries upholster the breach even as they upholster the untouched part, without paying more attention to it than to the rest of the wall. Caring nothing about the accident, they behave in the same way as on a non-gutted dwelling. The crevice is closed, in course of time, not intentionally, but solely by the action of the usual spinning.

We arrive at the same conclusion on the subject of the House Spider. Walking about her platform every night, she lays fresh courses without drawing a distinction between the solid and the

---

[1] The Processionaries are Moth-caterpillars that feed on various leaves and march in file, laying a silken trail as they go.—*Translator's Note*.

hollow. She has not deliberately put a patch in the torn texture; she has simply gone on with her ordinary business. If it happen that the hole is eventually closed, this fortunate result is the outcome not of a special purpose, but of an unvarying method of work.

Besides, it is evident that, if the Spider really wished to mend her web, all her endeavours would be concentrated upon the rent. She would devote to it all the silk at her disposal and obtain in one sitting a piece very like the rest of the web. Instead of that, what do we find? Almost nothing: a hardly visible gauze.

The thing is obvious: the Spider did on that rent what she did every elsewhere, neither more nor less. Far from squandering silk upon it, she saved her silk so as to have enough for the whole web. The gap will be better mended, little by little, afterwards, as the sheet is strengthened all over with new layers. And this will take long. Two months later, the window—my work—still shows through and makes a dark stain against the dead-white of the fabric.

Neither weavers nor spinners, therefore, know how to repair their work. Those wonderful manufacturers of silk-stuffs lack the least glimmer of that sacred lamp, reason, which enables the stupidest of darning-women to mend the heel of an old stocking. The office of inspector of Spiders' webs would have its uses, even if it merely succeeded in ridding us of a mistaken and mischievous idea.

CHAPTER XI

THE GARDEN SPIDERS: THE LIME-SNARE

THE spiral network of the Epeirae possesses contrivances of fearsome cunning. Let us give our attention by preference to that of the Banded Epeira or that of the Silky Epeira, both of which can be observed at early morning in all their freshness.

The thread that forms them is seen with the naked eye to differ from that of the framework and the spokes. It glitters in the sun, looks as though it were knotted and gives the impression of a chaplet of atoms. To examine it through the lens on the web itself is scarcely feasible, because of the shaking of the fabric, which trembles at the least breath. By passing a sheet of glass under the web and lifting it, I take away a few pieces of thread to study, pieces that remain fixed to the glass in parallel lines. Lens and microscope can now play their part.

The sight is perfectly astounding. Those threads, on the borderland between the visible and the invisible, are very closely twisted twine, similar to the gold cord of our officers' sword-knots. Moreover, they are hollow. The infinitely slender is a tube, a channel full of a viscous moisture resembling a strong solution of gum arabic. I can see a diaphanous trail of this moisture trickling through the broken ends. Under the pressure of the thin glass slide that covers them on the stage of the microscope, the twists lengthen out, become crinkled ribbons, traversed from end to end, through the middle, by a dark streak, which is the empty container.

The fluid contents must ooze slowly through the side of those tubular threads, rolled into twisted strings, and thus render the network sticky. It is sticky, in fact, and in such a way as to provoke surprise. I bring a fine straw flat down upon three or four rungs of a sector. However gentle the contact, adhesion is at once

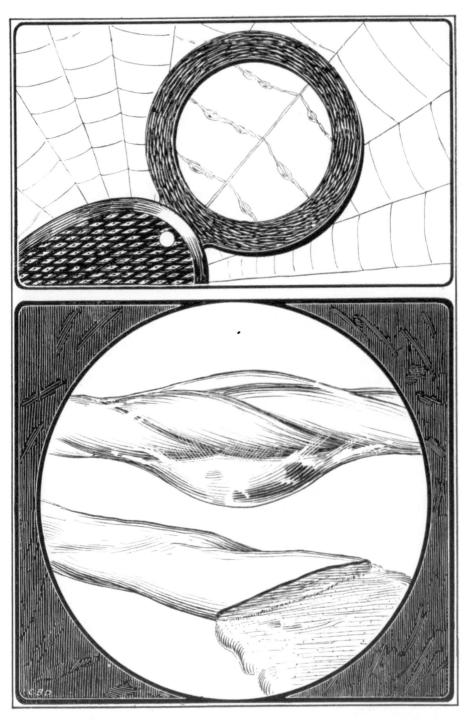

LENS AND MICROSCOPE NOW PLAY THEIR PART

established. When I lift the straw, the threads come with it and stretch to twice or three times their length, like a thread of India-rubber. At last, when over-taut, they loosen without breaking and resume their original form. They lengthen by unrolling their twist, they shorten by rolling it again; lastly, they become adhesive by taking the glaze of the gummy moisture wherewith they are filled.

In short, the spiral thread is a capillary tube finer than any that our physics will ever know. It is rolled into a twist so as to possess an elasticity that allows it, without breaking, to yield to the tugs of the captured prey; it holds a supply of sticky matter in reserve in its tube, so as to renew the adhesive properties of the surface by incessant exudation, as they become impaired by exposure to the air. It is simply marvellous.

The Epeira hunts not with springs, but with lime-snares. And such lime-snares! Everything is caught in them, down to the dandelion-plume that barely brushes against them. Nevertheless, the Epeira, who is in constant touch with her web, is not caught in them. Why?

Let us first of all remember that the Spider has contrived for herself, in the middle of her trap, a floor in whose construction the sticky spiral thread plays no part. We saw how this thread stops suddenly at some distance from the centre. There is here, covering a space which, in the larger webs, is about equal to the palm of one's hand, a fabric formed of spokes and of the commencement of the auxiliary spiral, a neutral fabric in which the exploring straw finds no adhesiveness anywhere.

Here, on this central resting-floor, and here only, the Epeira takes her stand, waiting whole days for the arrival of the game. However close, however prolonged her contact with this portion of the web, she runs no risk of sticking to it, because the gummy coating is lacking, as is the twisted and tubular structure, throughout the length of the spokes and throughout the extent of the auxiliary spiral. These pieces, together with the rest of the framework, are made of plain, straight, solid thread.

But, when a victim is caught, sometimes right at the edge of the web, the Spider has to rush up quickly, to bind it and overcome its attempts to free itself. She is walking then upon her network; and I do not find that she suffers the least inconvenience. The lime-threads are not even lifted by the movements of her legs.

In my boyhood, when a troop of us would go, on Thursdays,[1] to try and catch a Goldfinch in the hemp-fields, we used, before covering the twigs with glue, to grease our fingers with a few drops of oil, lest we should get them caught in the sticky matter. Does the Epeira know the secret of fatty substances? Let us try.

I rub my exploring straw with slightly oiled paper. When applied to the spiral thread of the web, it now no longer sticks to it. The principle is discovered. I pull out the leg of a live Epeira. Brought just as it is into contact with the lime-threads, it does not stick to them any more than to the neutral cords, whether spokes or parts of the framework. We were entitled to expect this, judging by the Spider's general immunity.

But here is something that wholly alters the result. I put the leg to soak for a quarter of an hour in disulphide of carbon, the best solvent of fatty matters. I wash it carefully with a brush dipped in the same fluid. When this washing is finished, the leg sticks to the snaring-thread quite easily and adheres to it just as well as anything else would, the unoiled straw, for instance.

Did I guess aright when I judged that it was a fatty substance that preserved the Epeira from the snares of her sticky Catherine-wheel? The action of the carbon disulphide seems to say yes. Besides, there is no reason why a substance of this kind, which plays so frequent a part in animal economy, should not coat the Spider very slightly by the mere act of perspiration. We used to rub our fingers with a little oil before handling the twigs in which the Goldfinch was to be caught; even so the Epeira varnishes herself with a special sweat, to operate on any part of her web without fear of the lime-threads.

However, an unduly protracted stay on the sticky threads would have its drawbacks. In the long run, continual contact with those threads might produce a certain adhesion and inconvenience the Spider, who must preserve all her agility in order to rush upon the prey before it can release itself. For this reason, gummy threads are never used in building the post of interminable waiting.

It is only on her resting-floor that the Epeira sits, motionless and with her eight legs outspread, ready to mark the least quiver in the

---

[1]The weekly half-holiday in French schools.—*Translator's Note*.

net. It is here, again, that she takes her meals, often long-drawn-out, when the joint is a substantial one; it is hither that, after trussing and nibbling it, she drags her prey at the end of a thread, to consume it at her ease on a non-viscous mat. As a hunting-post and refectory, the Epeira has contrived a central space, free from glue.

As for the glue itself, it is hardly possible to study its chemical properties, because the quantity is so slight. The microscope shows it trickling from the broken threads in the form of a transparent and more or less granular streak. The following experiment will tell us more about it.

With a sheet of glass passed across the web, I gather a series of lime-threads which remain fixed in parallel lines. I cover this sheet with a bell-jar standing in a depth of water. Soon, in this atmosphere saturated with humidity, the threads become enveloped in a watery sheath, which gradually increases and begins to flow. The twisted shape has by this time disappeared; and the channel of the thread reveals a chaplet of translucent orbs, that is to say, a series of extremely fine drops.

In twenty-four hours, the threads have lost their contents and are reduced to almost invisible streaks. If I then lay a drop of water on the glass, I get a sticky solution, similar to that which a particle of gum arabic might yield. The conclusion is evident: the Epeira's glue is a substance that absorbs moisture freely. In an atmosphere with a high degree of humidity, it becomes saturated and percolates by sweating through the side of the tubular threads.

These data explain certain facts relating to the work of the net. The full-grown Banded and Silky Epeirae weave at very early hours, long before dawn. Should the air turn misty, they sometimes leave that part of the task unfinished: they build the general framework, they lay the spokes, they even draw the auxiliary spiral, for all these parts are unaffected by excess of moisture; but they are very careful not to work at the lime-threads, which, if soaked by the fog, would dissolve into sticky shreds and lose their efficacy by being wetted. The net that was started will be finished to-morrow, if the atmosphere be favourable.

While the highly-absorbent character of the snaring-thread has its drawbacks, it also has compensating advantages. Both Epeirae, when hunting by day, affect those hot places, exposed to the fierce rays of the sun, wherein the Crickets delight. In the torrid heats of the dog-days, therefore, the lime-threads, but for special provisions, would be liable to dry up, to shrivel into stiff and lifeless filaments. But the very opposite happens. At the most scorching times of the day, they continue supple, elastic and more and more adhesive.

How is this brought about? By their very powers of absorption. The moisture of which the air is never deprived penetrates them slowly; it dilutes the thick contents of their tubes to the requisite degree and causes it to ooze through, as and when the earlier stickiness decreases. What bird-catcher could vie with the Garden Spider in the art of laying lime-snares? And all this industry and cunning for the capture of a Moth!

Then, too, what a passion for production! Knowing the diameter of the orb and the number of coils, we can easily calculate the total length of the sticky spiral. We find that, in one sitting, each time that she remakes her web, the Angular Epeira produces some twenty yards of gummy thread. The more skilful Silky Epeira produces thirty. Well, during two months, the Angular Epeira, my neighbour, renewed her snare nearly every evening. During that period, she manufactured something like three-quarters of a mile of this tubular thread, rolled into a tight twist and bulging with glue.

I should like an anatomist endowed with better implements than mine and with less tired eyesight to explain to us the work of the marvellous rope-yard. How is the silky matter moulded into a capillary tube? How is this tube filled with glue and tightly twisted? And how does this same wire-mill also turn out plain threads, wrought first into a framework and then into muslin and satin; next, a russet foam, such as fills the wallet of the Banded Epeira; next, the black stripes stretched in meridian curves on that same wallet? What a number of products to come from that curious factory, a Spider's belly! I behold the results, but fail to understand the working of the machine. I leave the problem to the masters of the microtome and the scalpel.

CHAPTER XII

THE GARDEN SPIDERS: THE TELEGRAPH-WIRE

Of the six Garden Spiders that form the object of my observations, two only, the Banded and the silky Epeira, remain constantly in their webs, even under the blinding rays of a fierce sun. The others, as a rule, do not show themselves until nightfall. At some distance from the net, they have a rough and ready retreat in the brambles, an ambush made of a few leaves held together by stretched threads. It is here that, for the most part, they remain in the daytime, motionless and sunk in meditation.

But the shrill light that vexes them is the joy of the fields. At such times, the Locust hops more nimbly than ever, more gaily skims the Dragon-fly. Besides, the limy web, despite the rents suffered during the night, is still in serviceable condition. If some giddy-pate allow himself to be caught, will the Spider, at the distance whereto she has retired, be unable to take advantage of the windfall? Never fear. She arrives in a flash. How is she apprised? Let us explain the matter.

The alarm is given by the vibration of the web, much more than by the sight of the captured object. A very simple experiment will prove this. I lay upon a Banded Epeira's lime-threads a Locust that second asphyxiated with carbon disulphide. The carcass is placed in front, or behind, or at either side of the Spider, who sits moveless in the centre of the net. If the test is to be applied to a species with a daytime hiding-place amid the foliage, the dead Locust is laid on the web, more or less near the centre, no matter how.

In both cases, nothing happens at first. The Epeira remains in her motionless attitude, even when the morsel is at a short distance in front of her. She is indifferent to the presence of the game, does not seem to perceive it, so much so that she ends by wearing out my patience. Then, with a long straw, which enables me to conceal myself slightly, I set the dead insect trembling.

That is quite enough. The Banded Epeira and the Silky Epeira hasten to the central floor; the others come down from the branch; all go to the Locust, swathe him with tape, treat him, in short, as they would treat a live prey captured under normal conditions. It took the shaking of the web to decide them to attack.

Perhaps the grey colour of the Locust is not sufficiently conspicuous to attract attention by itself. Then let us try red, the brightest colour to our retina and probably also to the Spiders'. None of the game hunted by the Epeirae being clad in scarlet, I make a small bundle out of red wool, a bait of the size of a Locust. I glue it to the web.

My stratagem succeeds. As long as the parcel is stationary, the Spider is not roused; but, the moment it trembles, stirred by my straw, she runs up eagerly.

There are silly ones who just touch the thing with their legs and, without further enquiries, swathe it in silk after the manner of the usual game. They even go so far as to dig their fangs into the bait, following the rule of the preliminary poisoning. Then and then only the mistake is recognized and the tricked Spider retires and does not come back, unless it be long afterwards, when she flings the cumbersome object out of the web.

There are also clever ones. Like the others, these hasten to the red-woollen lure, which my straw insidiously keeps moving; they come from their tent among the leaves as readily as from the centre of the web; they explore it with their palpi and their legs; but, soon perceiving that the thing is valueless, they are careful not to spend their silk on useless bonds. My quivering bait does not deceive them. It is flung out after a brief inspection.

Still, the clever ones, like the silly ones, run even from a distance, from their leafy ambush. How do they know? Certainly not by sight. Before recognizing their mistake, they have to hold the object between their legs and even to nibble at it a little. They are extremely short-sighted. At a hand's-breadth's distance, the lifeless prey, unable to shake the web, remains unperceived. Besides, in many cases, the hunting takes place in the dense darkness of the night, when sight, even if it were good, would not avail.

If the eyes are insufficient guides, even close at hand, how will it be when the prey has to be spied from afar! In that case, an intelligence-apparatus for long-distance work becomes indispensable. We have no difficulty in detecting the apparatus.

Let us look attentively behind the web of any Epeira with a daytime hiding-place: we shall see a thread that starts from the centre of the network, ascends in a slanting line outside the plane of the web and ends at the ambush where the Spider lurks all day. Except at the central point, there is no connection between this thread and the rest of the work, no interweaving with the scaffolding-threads. Free of impediment, the line runs straight from the centre of the net to the ambush-tent. Its length averages twenty-two inches. The Angular Epeira, settled high up in the trees, has shown me some as long as eight or nine feet.

There is no doubt that this slanting line is a foot-bridge which allows the Spider to repair hurriedly to the web, when summoned by urgent business, and then, when her round is finished, to return to her hut. In fact, it is the road which I see her follow, in going and coming. But is that all? No; for, if the Epeira had no aim in view but a means of rapid transit between her tent and the net, the foot-bridge would be fastened to the upper edge of the web. The journey would be shorter and the slope less steep.

Why, moreover, does this line always start in the centre of the sticky network and nowhere else? Because that is the point where the spokes meet and, therefore, the common centre of vibration. Anything that moves upon the web sets it shaking. All then that is needed is a thread issuing from this central point to convey to a distance the news of a prey struggling in some part or other of the net. The slanting cord, extending outside the plane of the web, is more than a foot-bridge: it is, above all, a signalling-apparatus, a telegraph-wire.

Let us try experiment. I place a Locust on the network. Caught in the sticky toils, he plunges about. Forthwith, the Spider issues impetuously from her hut, comes down the foot-bridge, makes a rush for the Locust, wraps him up and operates on him according to rule. Soon after, she hoists him, fastened by a line to her spinneret, and drags him to her hiding-place, where a long banquet will be held. So far, nothing new: things happen as usual.

I leave the Spider to mind her own affairs for some days, before I interfere with her. I again propose to give her a Locust; but, this time, I first cut the signalling-thread with a touch of the scissors, without shaking any part of the edifice. The game is then laid on the web. Complete success: the entangled insect struggles, sets the net quivering; the Spider, on her side, does not stir, as though heedless of events.

The idea might occur to one that, in this business, the Epeira stays motionless in her cabin since she is prevented from hurrying down, because the foot-bridge is broken. Let us undeceive ourselves: for one road open to her there are a hundred, all ready to bring her to the place where her presence is now required. The network is fastened to the branches by a host of lines, all of them very easy to cross. Well, the Epeira embarks upon none of them, but remains moveless and self-absorbed.

Why? Because her telegraph, being out of order, no longer tells her of the shaking of the web. The captured prey is too far off for her to see it; she is all unwitting. A good hour passes, with the Locust still kicking, the Spider impassive, myself watching. Nevertheless, in the end, the Epeira wakes up: no longer feeling the signalling-thread, broken by my scissors, as taut as usual under her legs, she comes to look into the state of things. The web is reached, without the least difficulty, by one of the lines of the framework, the first that offers. The Locust is then perceived and forthwith enswathed, after which the signalling-thread is remade, taking the place of the one which I have broken. Along this road the Spider goes home, dragging her prey behind her.

My neighbour, the mighty Angular Epeira, with her telegraph-wire nine feet long, has even better things in store for me. One morning, I find her web, which is now deserted, almost intact, a proof that the night's hunting has not been good. The animal must be hungry. With a piece of game for a bait, I hope to bring her down from her lofty retreat.

I entangle in the web a rare morsel, a Dragon-fly, who struggles desperately and sets the whole net a-shaking. The other, up above, leaves her lurking-place amid the cypress-foliage, strides swiftly down along her telegraph-wire, comes to the Dragon-fly, trusses her and at once climbs home again by the same road, with her prize dangling at her heels by a thread. The final sacrifice will take place in the quiet of the leafy sanctuary.

A few days later, I renew my experiment under the same conditions, but, this time, I first cut the signalling-thread. In vain I select a large Dragon-fly, a very restless prisoner; in vain I exert my patience: the Spider does not come down all day. Her telegraph being broken, she receives no notice of what is happening nine feet below. The entangled morsel remains where it lies, not despised, but unknown. At nightfall, the Epeira leaves her cabin, passes over the ruins of her web, finds the Dragon-fly and eats her on the spot, after which the net is renewed.

One of the Epeirae whom I have had the opportunity of examining simplifies the system, while retaining the essential mechanism of a transmission-thread. This is the Crater Epeira (*Epeira cratera*, WALCK.), a species seen in spring, at which time she indulges especially in the chase of the Domestic Bee, upon the flowering rosemaries. At the leafy end of a branch, she builds a sort of silken shell, the shape and size of an acorn-cup. This is where she sits, with her paunch contained in the round cavity and her forelegs resting on the ledge, ready to leap. The lazy creature loves this position and rarely stations herself head downwards on the web, as do the others. Cosily ensconced in the hollow of her cup, she awaits the approaching game.

Her web, which is vertical, as is the rule among the Epeirae, is of a fair size and always very near the bowl wherein the Spider takes her ease. Moreover, it touches the bowl by means of an angular extension; and the angle always contains one spoke which the Epeira, seated, so to speak, in her crater, has constantly under her legs. This spoke, springing from the common focus of the vibrations from all parts of the network, is eminently fitted to keep the Spider informed of whatsoever happens. It has a double office: it forms part of the Catherine-wheel supporting the lime-threads and it warns the Epeira by its vibrations. A special thread is here superfluous.

The other snarers, on the contrary, who occupy a distant retreat by day, cannot do without a private wire that keeps them in permanent communication with the deserted web. All of them have one, in point of fact, but only when age comes, age prone to rest and to long slumbers. In their youth, the Epeirae, who are then very wide-awake, know nothing of the art of telegraphy. Besides, their web, a short-lived work whereof hardly a trace remains on the morrow, does not allow of this kind of industry. It is no use going to the expense of a signalling-apparatus for a ruined snare wherein nothing can now be caught. Only the old Spiders, meditating or dozing in their green tent, are warned from afar, by telegraph, of what takes place on the web.

To save herself from keeping a close watch that would degenerate into drudgery and to remain alive to events even when resting, with her back turned on the net, the ambushed Spider always has her foot upon the telegraph-wire. Of my observations on this subject, let me relate the following, which will be sufficient for our purpose.

An Angular Epeira, with a remarkably fine belly, has spun her web between two laurestine-shrubs, covering a width of nearly a yard. The sun beats upon the snare, which is abandoned long before dawn. The Spider is in her day manor, a resort easily discovered by following the telegraph-wire. It is a vaulted chamber of dead leaves, joined together with a few bits of silk. The refuge is deep: the Spider disappears in it entirely, all but her rounded hind-quarters, which bar the entrance to the donjon.

With her front half plunged into the back of her hut, the Epeira certainly cannot see her web. Even if she had good sight, instead of being purblind, her position could not possibly allow her to keep the prey in view. Does she give up hunting during this period, of bright sunlight? Not at all. Look again.

Wonderful! One of her hind-legs is stretched outside the leafy cabin; and the signalling-thread ends just at the tip of that leg. Whoso has not seen the Epeira in this attitude, with her hand, so to speak, on the telegraph-receiver, knows nothing of one of the most curious instances of animal cleverness. Let any game appear upon the scene; and the slumberer, forthwith aroused by means of the leg receiving the vibrations, hastens up. A Locust whom I myself lay on the web procures her this agreeable shock and what follows. If she is satisfied with her bag, I am still more satisfied with what I have learnt.

The occasion is too good not to find out, under better conditions as regards approach, what the inhabitant of the cypress-trees has already shown me. The next morning, I cut the telegraph-wire, this time as long as one's arm and held, like yesterday, by one

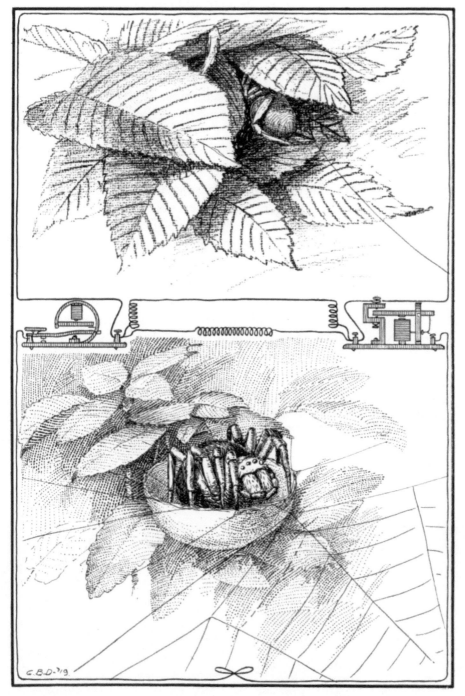

THE SPIDER IS IN HER DAY MANOR, A RESORT EASILY DISCOVERED BY FOLLOWING THE TELEGRAPH WIRE (UPPER) COSILY ENSCONCED IN THE HOLLOW OF HER CUP (LOWER)

of the hind-legs stretched outside the cabin. I then place on the web a double prey, a Dragon-fly and a Locust. The latter kicks out with his long, spurred shanks; the other flutters her wings. The web is tossed about to such an extent that a number of leaves, just beside the Epeira's nest, move, shaken by the threads of the framework affixed to them.

And this vibration, though so close at hand, does not rouse the Spider in the least, does not make her even turn round to enquire what is going on. The moment that her signalling-thread ceases to work, she knows nothing of passing events. All day long, she remains without stirring. In the evening, at eight o'clock, she sallies forth to weave the new web and at last finds the rich windfall whereof she was hitherto unaware.

One word more. The web is often shaken by the wind. The different parts of the framework, tossed and teased by the eddying air-currents, cannot fail to transmit their vibration to the signalling-thread. Nevertheless, the Spider does not quit her hut and remains indifferent to the commotion prevailing in the net. Her line, therefore, is something better than a bell-rope that pulls and communicates the impulse given: it is a telephone capable, like our own, of transmitting infinitesimal waves of sound. Clutching her telephone-wire with a toe, the Spider listens with her leg; she perceives the innermost vibrations; she distinguishes between the vibration proceeding from a prisoner and the mere shaking caused by the wind.

CHAPTER XIII

THE GARDEN SPIDERS: PAIRING AND HUNTING

NOTWITHSTANDING the importance of the subject, I shall not enlarge upon the nuptials of the Epeirae, grim natures whose loves easily turn to tragedy in the mystery of the night. I have but once been present at the pairing and for this curious experience I must thank my lucky star and my fat neighbour, the Angular Epeira, whom I visit so often by lantern-light. Here you have it.

It is the first week of August, at about nine o'clock in the evening, under a perfect sky, in calm, hot weather. The Spider has not yet constructed her web and is sitting motionless on her suspension-cable. The fact that she should be slacking like this, at a time when her building-operations ought to be in full swing, naturally astonishes me. Can something unusual be afoot?

Even so. I see hastening up from the neighbouring bushes and embarking on the cable a male, a dwarf, who is coming, the whipper-snapper, to pay his respects to the portly giantess. How has he, in his distant corner, heard of the presence of the nymph ripe for marriage? Among the Spiders, these things are learnt in the silence of the night, without a summons, without a signal, none knows how.

Once, the Great Peacock[1], apprised by the magic effluvia, used to come from miles around to visit the recluse in her bell-jar in my study. The dwarf of this evening, that other nocturnal pilgrim, crosses the intricate tangle of the branches without a mistake and makes straight for the rope-walker. He has as his guide the infallible compass that brings every Jack and his Jill together.

---

[1] Cf. *Social Life in the Insect World*, by J. H. Fabre, translated by Bernard Miall: chap. xiv.—*Translator's Note*.

He climbs the slope of the suspension-cord; he advances circumspectly, step by step. He stops some distance away, irresolute. Shall he go closer? Is this the right moment? No. The other lifts a limb and the scared visitor hurries down again. Recovering from his fright, he climbs up once more, draws a little nearer. More sudden flights, followed by fresh approaches, each time nigher than before. This restless running to and fro is the declaration of the enamoured swain.

Perseverance spells success. The pair are now face to face, she motionless and grave, he all excitement. With the tip of his leg, he ventures to touch the plump wench. He has gone too far, daring youth that he is! Panic-stricken, he takes a header, hanging by his safety-line. It is only for a moment, however. Up he comes again. He has learnt, from certain symptoms, that we are at last yielding to his blandishments.

With his legs and especially with his palpi, or feelers, he teases the buxom gossip, who answers with curious skips and bounds. Gripping a thread with her front tarsi, or fingers, she turns, one after the other, a number of back somersaults, like those of an acrobat on the trapeze. Having done this, she presents the underpart of her paunch to the dwarf and allows him to fumble at it a little with his feelers. Nothing more: it is done.

The object of the expedition is attained. The whipper-snapper makes off at full speed, as though he had the Furies at his heels. If he remained, he would presumably be eaten. These exercises on the tight-rope are not repeated. I kept watch in vain on the following evenings: I never saw the fellow again.

When he is gone, the bride descends from the cable, spins her web and assumes the hunting-attitude. We must eat to have silk, we must have silk to eat and especially to weave the expensive cocoon of the family. There is therefore no rest, not even after the excitement of being married.

The Epeirae are monuments of patience in their lime-snare. With her head down and her eight legs wide-spread, the Spider occupies the centre of the web, the receiving-point of the information sent along the spokes. If anywhere, behind or before, a vibration occur, the sign of a capture, the Epeira knows about it, even without the aid of sight. She hastens up at once.

PANIC-STRICKEN HE TAKES A HEADER

Until then, not a movement: one would think that the animal was hypnotized by her watching. At most, on the appearance of anything suspicious, she begins shaking her nest. This is her way of inspiring the intruder with awe. If I myself wish to provoke the singular alarm, I have but to tease the Epeira with a bit of straw. You cannot have a swing without an impulse of some sort. The terror-stricken Spider, who wishes to strike terror into others, has hit upon something much better. With nothing to push her, she swings with her floor of ropes. There is no effort, no visible exertion. Not a single part of the animal moves; and yet everything trembles. Violent shaking proceeds from apparent inertia. Rest causes commotion.

When calm is restored, she resumes her attitude, ceaselessly pondering the harsh problem of life:

'Shall I dine to-day, or not?'

Certain privileged beings, exempt from those anxieties, have food in abundance and need not struggle to obtain it. Such is the Gentle, who swims blissfully in the broth of the putrefying adder. Others—and, by a strange irony of fate, these are generally the most gifted—only manage to eat by dint of craft and patience.

You are of their company, O my industrious Epeirae! So that you may dine, you spend your treasures of patience nightly; and often without result. I sympathize with your woes, for I, who am as concerned as you about my daily bread, I also doggedly spread my net, the net for catching ideas, a more elusive and less substantial prize than the Moth. Let us not lose heart. The best part of life is not in the present, still less in the past; it lies in the future, the domain of hope. Let us wait.

All day long, the sky, of a uniform grey, has appeared to be brewing a storm. In spite of the threatened downpour, my neighbour, who is a shrewd weather-prophet, has come out of the cypress-tree and begun to renew her web at the regular hour. Her forecast is correct: it will be a fine night. See, the steaming-pan of the clouds splits open; and, through the apertures, the moon peeps, inquisitively. I too, lantern in hand, am peeping. A gust of wind from the north clears the realms on high; the sky becomes magnificent; perfect calm reigns below. The Moths begin their nightly rounds. Good! One is caught, a mighty fine one. The Spider will dine to-day.

What happens next, in an uncertain light, does not lend itself to accurate observation. It is better to turn to those Garden Spiders who never leave their web and who hunt mainly in the daytime. The Banded and the Silky Epeira, both of whom live on the rosemaries in the enclosure, shall show us in broad day-light the innermost details of the tragedy.

I myself place on the lime-snare a victim of my selecting. Its six legs are caught without more ado. If the insect raises one of its tarsi and pulls towards itself, the treacherous thread follows, unwinds slightly and, without letting go or breaking, yields to the captive's desperate jerks. Any limb released only tangles the others still more and is speedily recaptured by the sticky matter. There is no means of escape, except by smashing the trap with a sudden effort whereof even powerful insects are not always capable.

Warned by the shaking of the net, the Epeira hastens up; she turns round about the quarry; she inspects it at a distance, so as to ascertain the extent of the danger before attacking. The strength of the snareling will decide the plan of campaign. Let us first suppose the usual case, that of an average head of game, a Moth or Fly of some sort. Facing her prisoner, the Spider contracts her abdomen slightly and touches the insect for a moment with the end of her spinnerets; then, with her front tarsi, she sets her victim spinning. The Squirrel, in the moving cylinder of his cage, does not display a more graceful or nimbler dexterity. A cross-bar of the sticky spiral serves as an axis for the tiny machine, which turns, turns swiftly, like a spit. It is a treat to the eyes to see it revolve.

What is the object of this circular motion? See, the brief contact of the spinnerets has given a starting-point for a thread, which the Spider must now draw from her silk-warehouse and gradually roll around the captive, so as to swathe him in a winding-sheet which will overpower any effort made. It is the exact process employed in our wire-mills: a motor-driven spool revolves and, by its action, draws the wire through the narrow eyelet of a steel plate, making it of the fineness required, and, with the same movement, winds it round and round its collar.

Even so with the Epeira's work. The Spider's front tarsi are the motor; the revolving spool is the captured insect; the steel eyelet is the aperture of the spinnerets. To bind the subject with precision and dispatch nothing could be better than this inexpensive and highly-effective method.

Less frequently, a second process is employed. With a quick movement, the Spider herself turns round about the motionless insect, crossing the web first at the top and then at the bottom and gradually placing the fastenings of her line. The great elasticity of the lime-threads allows the Epeira to fling herself time after time right into the web and to pass through it without damaging the net.

Let us now suppose the case of some dangerous game: a Praying Mantis, for instance, brandishing her lethal limbs, each hooked and fitted with a double saw; an angry Hornet, darting her awful sting; a sturdy Beetle, invincible under his horny armour. These are exceptional morsels, hardly ever known to the Epeirae. Will they be accepted, if supplied by my stratagems?

They are, but not without caution. The game is seen to be perilous of approach and the Spider turns her back upon it, instead of facing it; she trains her rope-cannon upon it. Quickly, the hind-legs draw from the spinnerets something much better than single cords. The whole silk-battery works at one and the same time, firing a regular volley of ribbons and sheets, which a wide movement of the legs spreads fan-wise and flings over the entangled prisoner. Guarding against sudden starts, the Epeira casts her armfuls of bands on the front-and hind-parts, over the legs and over the wings, here, there and everywhere, extravagantly. The most fiery prey is promptly mastered under this avalanche. In vain, the Mantis tries to open her saw-toothed arm-guards; in vain, the Hornet makes play with her dagger; in vain, the Beetle stiffens his legs and arches his back: a fresh wave of threads swoops down and paralyses every effort.

These lavished, far-flung ribbons threaten to exhaust the factory; it would be much more economical to resort to the method of the spool; but, to turn the machine, the Spider would have to go up to it and work it with her leg. This is too risky; and hence the continuous spray of silk, at a safe distance. When all is used up, there is more to come.

Still, the Epeira seems concerned at this excessive outlay. When circumstances permit, she gladly returns to the mechanism of the revolving spool. I saw her practise this abrupt change of tactics on a big Beetle, with a smooth, plump body, which lent itself admirably to the rotary process. After depriving the beast of all power of movement, she went up to it and turned her corpulent victim as she would have done with a medium-sized Moth.

But with the Praying Mantis, sticking out her long legs and her spreading wings, rotation is no longer feasible. Then, until the quarry is thoroughly subdued, the spray of bandages goes on continuously, even to the point of drying up the silk-glands. A capture of this kind is ruinous. It is true that, except when I interfered, I have never seen the Spider tackle that formidable provender.

Be it feeble or strong, the game is now neatly trussed, by one of the two methods. The next move never varies. The bound insect is bitten, without persistency and without any wound that shows. The Spider next retires and allows the bite to act, which it soon does. She then returns.

If the victim be small, a Clothes-moth, for instance, it is consumed on the spot, at the place where it was captured. But, for a prize of some importance, on which she hopes to feast for many an hour, sometimes for many a day, the Spider needs a sequestered dining-room, where there is naught to fear from the stickiness of the network. Before going to it, she first makes her prey turn in the converse direction to that of the original rotation. Her object is to free the nearest spokes, which supplied pivots for the machinery. They are essential factors which it behoves her to keep intact, if need be by sacrificing a few cross-bars.

It is done; the twisted ends are put back into position. The well-trussed game is at last removed from the web and fastened on behind with a thread. The Spider then marches in front and the load is trundled across the web and hoisted to the resting-floor, which is both an inspection-post and a dining-hall. When the Spider is of a species that shuns the light and possesses a telegraph-line, she mounts to her daytime hiding-place along this line, with the game bumping against her heels.

While she is refreshing herself, let us enquire into the effects of the little bite previously administered to the silk-swathed captive. Does the Spider kill the patient with a view to avoiding unseasonable jerks, protests so disagreeable at dinner-time? Several reasons make me doubt it. In the first place, the attack is so much veiled as to have all the appearance of a mere kiss. Besides, it is made anywhere, at the first spot that offers. The expert slayers[1] employ methods of the highest precision: they give a stab in the neck, or under the throat; they wound the cervical nerve-centres, the seat of energy. The paralyzers, those accomplished anatomists, poison the motor nerve-centres, of which they know the number and position. The Epeira possesses none of this fearsome knowledge. She inserts her fangs at random, as the Bee does her sting. She does not select one spot rather than another; she bites indifferently at whatever comes within reach. This being so, her poison would have to possess unparalleled virulence to produce a corpse-like inertia no matter which the point attacked. I can scarcely believe in instantaneous death resulting from the bite, especially in the case of insects, with their highly-resistant organisms.

Besides, is it really a corpse that the Epeira wants, she who feeds on blood much more than on flesh? It were to her advantage to suck a live body, wherein the flow of the liquids, set in movement by the pulsation of the dorsal vessel, that rudimentary heart of insects, must act more freely than in a lifeless body, with its stagnant fluids. The game which the Spider means to suck dry might very well not be dead. This is easily ascertained.

I place some Locusts of different species on the webs in my menagerie, one on this, another on that. The Spider comes rushing up, binds the prey, nibbles at it gently and withdraws, waiting for the bite to take effect. I then take the insect and carefully strip it of its silken shroud. The Locust is not dead, far from it; one would even think that he had suffered no harm. I examine the released prisoner through the lens in vain; I can see no trace of a wound.

Can he be unscathed, in spite of the sort of kiss which I saw given to him just now? You would be ready to say so, judging by the furious way in which he kicks in my fingers. Nevertheless, when put on the ground, he walks awkwardly, he seems reluctant to hop. Perhaps it is a temporary trouble, caused by his terrible excitement in the web. It looks as though it would soon pass.

---

[1] Cf. Insect Life, by J. H. Fabre, translated by the author of Mademoiselle Mori: Chap. V.—*Translator's Note*.

I lodge my Locusts in cages, with a lettuce-leaf to console them for their trials; but they will not be comforted. A day elapses, followed by a second. Not one of them touches the leaf of salad; their appetite has disappeared. Their movements become more uncertain, as though hampered by irresistible torpor. On the second day, they are dead, every one irrecoverably dead.

The Epeira, therefore, does not incontinently kill her prey with her delicate bite; she poisons it so as to produce a gradual weakness, which gives the blood-sucker ample time to drain her victim, without the least risk, before the rigor mortis stops the flow of moisture.

The meal lasts quite twenty-four hours, if the joint be large; and to the very end the butchered insect retains a remnant of life, a favourable condition for the exhausting of the juices. Once again, we see a skilful method of slaughter, very different from the tactics in use among the expert paralyzers or slayers. Here there is no display of anatomical science. Unacquainted with the patient's structure, the Spider stabs at random. The virulence of the poison does the rest.

There are, however, some very few cases in which the bite is speedily mortal. My notes speak of an Angular Epeira grappling with the largest Dragon-fly in my district (*AEshna grandis*, Lin.). I myself had entangled in the web this head of big game, which is not often captured by the Epeirae. The net shakes violently, seems bound to break its moorings.

The Spider rushes from her leafy villa, runs boldly up to the giantess, flings a single bundle of ropes at her and, without further precautions, grips her with her legs, tries to subdue her and then digs her fangs into the Dragon-fly's back. The bite is prolonged in such a way as to astonish me. This is not the perfunctory kiss with which I am already familiar; it is a deep, determined wound. After striking her blow, the Spider retires to a certain distance and waits for her poison to take effect.

I at once remove the Dragon-fly. She is dead, really and truly dead. Laid upon my table and left alone for twenty-four hours, she makes not the slightest movement. A prick of which my lens cannot see the marks, so sharp-pointed are the Epeira's weapons, was enough, with a little insistence, to kill the powerful animal.

Proportionately, the Rattlesnake, the Horned Viper, the Trigonocephalus and other ill-famed serpents produce less paralysing effects upon their victims.

And these Epeirae, so terrible to insects, I am able to handle without any fear. My skin does not suit them. If I persuaded them to bite me, what would happen to me? Hardly anything. We have more cause to dread the sting of a nettle than the dagger which is fatal to Dragon-flies. The same virus acts differently upon this organism and that, is formidable here and quite mild there. What kills the insect may easily be harmless to us. Let us not, however, generalize too far. The Narbonne Lycosa, that other enthusiastic insect-huntress, would make us pay clearly if we attempted to take liberties with her.

It is not uninteresting to watch the Epeira at dinner. I light upon one, the Banded Epeira, at the moment, about three o'clock in the afternoon, when she has captured a Locust. Planted in the centre of the web, on her resting-floor, she attacks the venison at the joint of a haunch. There is no movement, not even of the mouth-parts, as far as I am able to discover. The mouth lingers, close-applied, at the point originally bitten. There are no intermittent mouthfuls, with the mandibles moving backwards and forwards. It is a sort of continuous kiss.

I visit my Epeira at intervals. The mouth does not change its place. I visit her for the last time at nine o'clock in the evening. Matters stand exactly as they did: after six hours' consumption, the mouth is still sucking at the lower end of the right haunch. The fluid contents of the victim are transferred to the ogress' belly, I know not how.

Next morning, the Spider is still at table. I take away her dish. Naught remains of the Locust but his skin, hardly altered in shape, but utterly drained and perforated in several places. The method, therefore, was changed during the night. To extract the non-fluent residue, the viscera and muscles, the stiff cuticle had to be tapped here, there and elsewhere, after which the tattered husk, placed bodily in the press of the mandibles, would have been chewed, rechewed and finally reduced to a pill, which the sated Spider throws up. This would have been the end of the victim, had I not taken it away before the time.

Whether she wound or kill, the Epeira bites her captive somewhere or other, no matter where. This is an excellent method on her part, because of the variety of the game that comes her way. I see her accepting with equal readiness whatever chance may send her: Butterflies and Dragon-flies, Flies and Wasps, small Dung-beetles and Locusts. If I offer her a Mantis, a Bumble-bee, an Anoxia—the equivalent of the common Cockchafer—and other dishes probably unknown to her race, she accepts all and any, large and small, thin-skinned and horny-skinned, that which goes afoot and that which takes winged flight. She is omnivorous, she preys on everything, down to her own kind, should the occasion offer.

Had she to operate according to individual structure, she would need an anatomical dictionary; and instinct is essentially unfamiliar with generalities: its knowledge is always confined to limited points. The Cerceres know their Weevils and their Buprestis-beetles absolutely; the Sphex their Grasshoppers, their Crickets and their Locusts; the Scoliae[1] their Cetonia- and Oryctes-grubs. Even so the other paralyzers. Each has her own victim and knows nothing of any of the others.

The same exclusive tastes prevail among the slayers. Let us remember, in this connection, *Philanthus apivorus*[2] and, especially, the Thomisus, the comely Spider who cuts Bees' throats. They understand the fatal blow, either in the neck or under the chin, a thing which the Epeira does not understand; but, just because of this talent, they are specialists. Their province is the Domestic Bee.

Animals are a little like ourselves: they excel in an art only on condition of specializing in it. The Epeira, who, being omnivorous, is obliged to generalize, abandons scientific methods and makes up for this by distilling a poison capable of producing torpor and even death, no matter what the point attacked.

Recognizing the large variety of game, we wonder how the Epeira manages not to hesitate amid those many diverse forms, how, for instance, she passes from the Locust to the Butterfly, so different

[1] The Scolia is a Digger-wasp, like the Cerceris and the Sphex, and feeds her larvae on the grubs of the Cetonia, or Rose-chafer, and the Oryctes, or Rhinoceros Beetle. Cf. The Life and Love of the Insect, by J. Henri Fabre, translated by Alexander Teixeira de Mattos: Chap. XI.—*Translator's Note*.

[2] Cf. Social Life in the Insect World, by J. H. Fabre, translated by Bernard Miall. chap. xiii., in which the name is given, by a printer's error, as Philanthus aviporus.—*Translator's Note*.

in appearance. To attribute to her as a guide an extensive zoological knowledge were wildly in excess of what we may reasonably expect of her poor intelligence. The thing moves, therefore it is worth catching: this formula seems to sum up the Spider's wisdom.

CHAPTER XIV

THE GARDEN SPIDERS: THE QUESTION OF PROPERTY

A DOG has found a bone. He lies in the shade, holding it between his paws, and studies it fondly. It is his sacred property, his chattel. An Epeira has woven her web. Here again is property; and owning a better title than the other. Favoured by chance and assisted by his scent, the Dog has merely had a find; he has neither worked nor paid for it. The Spider is more than a casual owner, she has created what is hers. Its substance issued from her body, its structure from her brain. If ever property was sacrosanct, hers is.

Far higher stands the work of the weaver of ideas, who tissues a book, that other Spider's web, and out of his thought makes something that shall instruct or thrill us. To protect our 'bone,' we have the police, invented for the express purpose. To protect the book, we have none but farcical means. Place a few bricks one atop the other; join them with mortar; and the law will defend your wall. Build up in writing an edifice of your thoughts; and it will be open to any one, without serious impediment, to abstract stones from it, even to take the whole, if it suit him. A rabbit-hutch is property; the work of the mind is not. If the animal has eccentric views as regards the possessions of others, we have ours as well.

'Might always has the best of the argument,' said La Fontaine, to the great scandal of the peace-lovers. The exigencies of verse, rhyme and rhythm, carried the worthy fabulist further than he intended: he meant to say that, in a fight between mastiffs and in other brute conflicts, the stronger is left master of the bone. He well knew that, as things go, success is no certificate of excellence. Others came, the notorious evil-doers of humanity, who made a law of the savage maxim that might is right.

We are the larvae with the changing skins, the ugly caterpillars of a society that is slowly, very slowly, wending its way to the triumph of right over might. When will this sublime metamorphosis be accomplished? To free ourselves from those wild-beast brutalities, must we wait for the ocean-plains of the southern hemisphere to flow to our side, changing the face of continents and renewing the glacial period of the Reindeer and the Mammoth? Perhaps, so slow is moral progress.

True, we have the bicycle, the motor-car, the dirigible airship and other marvellous means of breaking our bones; but our morality is not one rung the higher for it all. One would even say that, the farther we proceed in our conquest of matter, the more our morality recedes. The most advanced of our inventions consists in bringing men down with grapeshot and explosives with the swiftness of the reaper mowing the corn.

Would we see this might triumphant in all its beauty? Let us spend a few weeks in the Epeira's company. She is the owner of a web, her work, her most lawful property. The question at once presents itself: Does the Spider possibly recognize her fabric by certain trademarks and distinguish it from that of her fellows?

I bring about a change of webs between two neighbouring Banded Epeirae. No sooner is either placed upon the strange net than she makes for the central floor, settles herself head downwards and does not stir from it, satisfied with her neighbour's web as with her own. Neither by day nor by night does she try to shift her quarters and restore matters to their pristine state. Both Spiders think themselves in their own domain. The two pieces of work are so much alike that I almost expected this.

I then decide to effect an exchange of webs between two different species. I move the Banded Epeira to the net of the Silky Epeira and vice versa. The two webs are now dissimilar; the Silky Epeira's has a limy spiral consisting of closer and more numerous circles. What will the Spiders do, when thus put to the test of the unknown? One would think that, when one of them found meshes too wide for her under her feet, the other meshes too narrow, they would be frightened by this sudden change and decamp in terror. Not at all. Without a sign of perturbation, they remain, plant themselves in the centre and await the coming of the game, as though nothing extraordinary had happened. They do more than this. Days pass and, as long as the unfamiliar web is not wrecked to the extent of being unserviceable, they make no attempt to weave another in their own style. The Spider, therefore, is incapable of recognizing her web. She takes another's work for hers, even when it is produced by a stranger to her race.

We now come to the tragic side of this confusion. Wishing to have subjects for study within my daily reach and to save myself the trouble of casual excursions, I collect different Epeirae whom I find in the course of my walks and establish them on the shrubs in my enclosure. In this way, a rosemary-hedge, sheltered from the wind and facing the sun, is turned into a well-stocked menagerie. I take the Spiders from the paper bags wherein I had put them separately, to carry them, and place them on the leaves, with no further precaution. It is for them to make themselves at home. As a rule, they do not budge all day from the place where I put them: they wait for nightfall before seeking a suitable site whereon to weave a net.

Some among them show less patience. A little while ago, they possessed a web, between the reeds of a brook or in the holm-oak copses; and now they have none. They go off in search, to recover their property or seize on some one else's: it is all the same to them. I come upon a Banded Epeira, newly imported, making for the web of a Silky Epeira who has been my guest for some days now. The owner is at her post, in the centre of the net. She awaits the stranger with seeming impassiveness. Then suddenly they grip each other; and a desperate fight begins. The Silky Epeira is worsted. The other swathes her in bonds, drags her to the non-limy central floor and, in the calmest fashion, eats her. The dead Spider is munched for twenty-four hours and drained to the last drop, when the corpse, a wretched, crumpled ball, is at last flung aside. The web so foully conquered becomes the property of the stranger, who uses it, if it have not suffered too much in the contest.

There is here a shadow of an excuse. The two Spiders were of different species; and the struggle for life often leads to these exterminations among such as are not akin. What would happen if the two belonged to the same species? It is easily seen. I cannot rely upon spontaneous invasions, which may be rare under normal conditions, and I myself place a Banded Epeira on her kinswoman's web. A furious attack is made forthwith. Victory, after hanging for a moment in the balance, is once again decided in the stranger's favour. The vanquished party, this time a sister, is eaten without the slightest scruple. Her web becomes the property of the victor.

There it is, in all its horror, the right of might: to eat one's like and take away their goods. Man did the same in days of old: he stripped and ate his fellows. We continue to rob one another, both as nations and as individuals; but we no longer eat one another: the custom has grown obsolete since we discovered an acceptable substitute in the mutton-chop.

Let us not, however, blacken the Spider beyond her deserts. She does not live by warring on her kith and kin; she does not of her own accord attempt the conquest of another's property. It needs extraordinary circumstances to rouse her to these villainies. I take her from her web and place her on another's. From that moment, she knows no distinction between *meum* and *tuum*: the thing which the leg touches at once becomes real estate. And the intruder, if she be the stronger, ends by eating the occupier, a radical means of cutting short disputes.

Apart from disturbances similar to those provoked by myself, disturbances that are possible in the everlasting conflict of events, the Spider, jealous of her own web, seems to respect the webs of others. She never indulges in brigandage against her fellows except when dispossessed of her net, especially in the daytime, for weaving is never done by day: this work is reserved for the night. When, however, she is deprived of her livelihood and feels herself the stronger, then she attacks her neighbour, rips her open, feeds on her and takes possession of her goods. Let us make allowances and proceed.

We will now examine Spiders of more alien habits. The Banded and the Silky Epeira differ greatly in form and colouring. The first has a plump, olive-shaped belly, richly belted with white, bright-yellow and black; the second's abdomen is flat, of a silky white and pinked into festoons. Judging only by dress and figure, we should not think of closely connecting the two Spiders.

But high above shapes tower tendencies, those main characteristics which our methods of classification, so particular about minute details of form, ought to consult more widely than they do. The two dissimilar Spiders have exactly similar ways of living. Both of them prefer to hunt by day and never leave their webs; both sign their work with a zigzag flourish. Their nets are almost identical, so much so that the Banded Epeira uses the Silky Epeira's web after eating its owner. The Silky Epeira, on her side, when she is the stronger, dispossesses her belted cousin and devours her. Each is at home on the other's web, when the argument of might triumphant has ended the discussion.

Let us next take the case of the Cross Spider, a hairy beast of varying shades of reddish-brown. She has three large white spots upon her back, forming a triple-barred cross. She hunts mostly at night, shuns the sun and lives by day on the adjacent shrubs, in a shady retreat which communicates with the lime-snare by means of a telegraph-wire. Her web is very similar in structure and appearance to those of the two others. What will happen if I procure her the visit of a Banded Epeira?

The lady of the triple cross is invaded by day, in the full light of the sun, thanks to my mischievous intermediary. The web is deserted; the proprietress is in her leafy hut. The telegraph-wire performs its office; the Cross Spider hastens down, strides all round her property, beholds the danger and hurriedly returns to her hiding-place, without taking any measures against the intruder.

The latter, on her side, does not seem to be enjoying herself. Were she placed on the web of one of her sisters, or even on that of the Silky Epeira, she would have posted herself in the centre, as soon as the struggle had ended in the other's death. This time there is no struggle, for the web is deserted; nothing prevents her from taking her position in the centre, the chief strategic point; and yet she does not move from the place where I put her.

I tickle her gently with the tip of a long straw. When at home, if teased in this way, the Banded Epeira—like the others, for that matter—violently shakes the web to intimidate the aggressor. This time, nothing happens: despite my repeated enticements, the Spider does not stir a limb. It is as though she were numbed with terror. And she has reason to be: the other is watching her from her lofty loop-hole.

THE CROSS SPIDER HASTENS DOWN

This is probably not the only cause of her fright. When my straw does induce her to take a few steps, I see her lift her legs with some difficulty. She tugs a bit, drags her tarsi till she almost breaks the supporting threads. It is not the progress of an agile rope-walker; it is the hesitating gait of entangled feet. Perhaps the lime-threads are stickier than in her own web. The glue is of a different quality; and her sandals are not greased to the extent which the new degree of adhesiveness would demand.

Anyhow, things remain as they are for long hours on end: the Banded Epeira motionless on the edge of the web; the other lurking in her hut; both apparently most uneasy. At sunset, the lover of darkness plucks up courage. She descends from her green tent and, without troubling about the stranger, goes straight to the centre of the web, where the telegraph-wire brings her. Panic-stricken at this apparition, the Banded Epeira releases herself with a jerk and disappears in the rosemary-thicket.

The experiment, though repeatedly renewed with different subjects, gave me no other results. Distrustful of a web dissimilar to her own, if not in structure, at least in stickiness, the bold Banded Epeira shows the white feather and refuses to attack the Cross Spider. The latter, on her side, either does not budge from her day shelter in the foliage, or else rushes back to it, after taking a hurried glance at the stranger. She here awaits the coming of the night. Under favour of the darkness, which gives her fresh courage and activity, she reappears upon the scene and puts the intruder to flight by her mere presence, aided, if need be, by a cuff or two. Injured right is the victor.

Morality is satisfied; but let us not congratulate the Spider therefore. If the invader respects the invaded, it is because very serious reasons impel her. First, she would have to contend with an adversary ensconced in a stronghold whose ambushes are unknown to the assailant. Secondly, the web, if conquered, would be inconvenient to use, because of the lime-threads, possessing a different degree of stickiness from those which she knows so well. To risk one's skin for a thing of doubtful value were twice foolish. The Spider knows this and forbears.

But let the Banded Epeira, deprived of her web, come upon that of one of her kind or of the Silky Epeira, who works her gummy twine in the same manner: then discretion is thrown to the winds; the owner is fiercely ripped open and possession taken of the property.

Might is right, says the beast; or, rather, it knows no right. The animal world is a rout of appetites, acknowledging no other rein than impotence. Mankind, alone capable of emerging from the slough of the instincts, is bringing equity into being, is creating it slowly as its conception grows clearer. Out of the sacred rushlight, so flickering as yet, but gaining strength from age to age, man will make a flaming torch that will put an end, among us, to the principles of the brutes and, one day, utterly change the face of society.

CHAPTER XV

THE LABYRINTH SPIDER

WHILE the Epeirae, with their gorgeous net-tapestries, are incomparable weavers, many other Spiders excel in ingenious devices for filling their stomachs and leaving a lineage behind them: the two primary laws of living things. Some of them are celebrities of long-standing renown, who are mentioned in all the books.

Certain Mygales[1] inhabit a burrow, like the Narbonne Lycosa, but of a perfection unknown to the brutal Spider of the wastelands. The Lycosa surrounds the mouth of her shaft with a simple parapet, a mere collection of tiny pebbles, sticks and silk; the others fix a movable door to theirs, a round shutter with a hinge, a groove and a set of bolts. When the Mygale comes home, the lid drops into the groove and fits so exactly that there is no possibility of distinguishing the join. If the aggressor persist and seek to raise the trap-door, the recluse pushes the bolt, that is to say, plants her claws into certain holes on the opposite side to the hinge, props herself against the wall and holds the door firmly.

Another, the Argyroneta, or Water Spider, builds herself an elegant silken diving-bell, in which she stores air. Thus supplied with the wherewithal to breathe, she awaits the coming of the game and keeps herself cool meanwhile. At times of scorching heat, hers must be a regular sybaritic abode, such as eccentric man has sometimes ventured to build under water, with mighty blocks of stone and marble. The submarine palaces of Tiberius are no more than an odious memory; the Water Spider's dainty cupola still flourishes.

---

[1] Or Bird Spiders, known also as the American Tarantula.—*Translator's Note*.

If I possessed documents derived from personal observation, I should like to speak of these ingenious workers; I would gladly add a few unpublished facts to their life-history. But I must abandon the idea. The Water Spider is not found in my district. The Mygale, the expert in hinged doors, is found there, but very seldom. I saw one once, on the edge of a path skirting a copse. Opportunity, as we know, is fleeting. The observer, more than any other, is obliged to take it by the forelock. Preoccupied as I was with other researches, I but gave a glance at the magnificent subject which good fortune offered. The opportunity fled and has never returned.

Let us make up for it with trivial things of frequent encounter, a condition favourable to consecutive study. What is common is not necessarily unimportant. Give it our sustained attention and we shall discover in it merits which our former ignorance prevented us from seeing. When patiently entreated, the least of creatures adds its note to the harmonies of life.

In the fields around, traversed, in these days, with a tired step, but still vigilantly explored, I find nothing so often as the Labyrinth Spider (*Agelena labyrinthica*, CLERCK.). Not a hedge but shelters a few at its foot, amidst grass, in quiet, sunny nooks. In the open country and especially in hilly places laid bare by the wood-man's axe, the favourite sites are tufts of bracken, rock-rose, lavender, everlasting and rosemary cropped close by the teeth of the flocks. This is where I resort, as the isolation and kindliness of the supports lend themselves to proceedings which might not be tolerated by the unfriendly hedge.

Several times a week, in July, I go to study my Spiders on the spot, at an early hour, before the sun beats fiercely on one's neck. The children accompany me, each provided with an orange wherewith to slake the thirst that will not be slow in coming. They lend me their good eyes and supple limbs. The expedition promises to be fruitful.

We soon discover high silk buildings, betrayed at a distance by the glittering threads which the dawn has converted into dewy rosaries. The children are wonderstruck at those glorious chandeliers, so much so that they forget their oranges for a moment. Nor am I, on my part, indifferent. A splendid spectacle indeed is that of our Spider's labyrinth, heavy with the tears of the night and

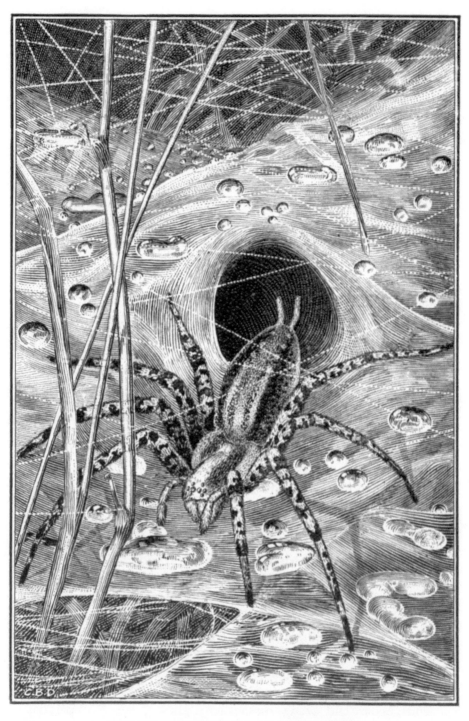

HEAVY WITH TEARS OF THE NIGHT

lit up by the first rays of the sun. Accompanied as it is by the Thrushes' symphony, this alone is worth getting up for.

Half an hour's heat; and the magic jewels disappear with the dew. Now is the moment to inspect the webs. Here is one spreading its sheet over a large cluster of rock-roses; it is the size of a handkerchief. A profusion of guy-ropes, attached to any chance projection, moor it to the brushwood. There is not a twig but supplies a contact-point. Entwined on every side, surrounded and surmounted, the bush disappears from view, veiled in white muslin.

The web is flat at the edges, as far as the unevenness of the support permits, and gradually hollows into a crater, not unlike the bell of a hunting-horn. The central portion is a cone-shaped gulf, a funnel whose neck, narrowing by degrees, dives perpendicularly into the leafy thicket to a depth of eight or nine inches.

At the entrance to the tube, in the gloom of that murderous alley, sits the Spider, who looks at us and betrays no great excitement at our presence. She is grey, modestly adorned on the thorax with two black ribbons and on the abdomen with two stripes in which white specks alternate with brown. At the tip of the belly, two small, mobile appendages form a sort of tail, a rather curious feature in a Spider.

The crater-shaped web is not of the same structure throughout. At the borders, it is a gossamer weft of sparse threads; nearer the centre, the texture becomes first fine muslin and then satin; lower still, on the narrower part of the opening, it is a network of roughly lozenged meshes. Lastly, the neck of the funnel, the usual resting-place, is formed of solid silk.

The Spider never ceases working at her carpet, which represents her investigation-platform. Every night she goes to it, walks over it, inspecting her snares, extending her domain and increasing it with new threads. The work is done with the silk constantly hanging from the spinnerets and constantly extracted as the animal moves about. The neck of the funnel, being more often walked upon than the rest of the dwelling, is therefore provided with a thicker upholstery. Beyond it are the slopes of the crater, which are also much-frequented regions. Spokes of some regularity fix the diameter of the mouth; a swaying walk and the guiding aid of the caudal appendages have laid lozengy meshes across these spokes. This part has been strengthened by the nightly rounds of inspection. Lastly come the less-visited expanses, which consequently have a thinner carpet.

At the bottom of the passage dipping into the brushwood, we might expect to find a secret cabin, a wadded cell where the Spider would take refuge in her hours of leisure. The reality is something entirely different. The long funnel-neck gapes at its lower end, where a private door stands always ajar, allowing the animal, when hard-pushed, to escape through the grass and gain the open.

It is well to know this arrangement of the home, if you wish to capture the Spider without hurting her. When attacked from the front, the fugitive runs down and slips through the postern-gate at the bottom. To look for her by rummaging in the brushwood often leads to nothing, so swift is her flight; besides, a blind search entails a great risk of maiming her. Let us eschew violence, which is but seldom successful, and resort to craft.

We catch sight of the Spider at the entrance to her tube. If practicable, squeeze the bottom of the tuft, containing the neck of the funnel, with both hands. That is enough; the animal is caught. Feeling its retreat cut off, it readily darts into the paper bag held out to it; if necessary, it can be stimulated with a bit of straw. In this way, I fill my cages with subjects that have not been demoralized by contusions.

The surface of the crater is not exactly a snare. It is just possible for the casual pedestrian to catch his legs in the silky carpets; but giddy-pates who come here for a walk must be very rare. What is wanted is a trap capable of securing the game that hops or flies. The Epeira has her treacherous limed net; the Spider of the bushes has her no less treacherous labyrinth.

Look above the web. What a forest of ropes! It might be the rigging of a ship disabled by a storm. They run from every twig of the supporting shrubs, they are fastened to the tip of every branch. There are long ropes and short ropes, upright and slanting, straight and bent, taut and slack, all criss-cross and a-tangle, to the height of three feet or so in inextricable disorder. The whole forms a chaos of netting, a labyrinth which none can pass through, unless he be endowed with wings of exceptional power.

We have here nothing similar to the lime-threads used by the Garden Spiders. The threads are not sticky; they act only by their confused multitude. Would you care to see the trap at work? Throw a small Locust into the rigging. Unable to obtain a steady foothold on that shaky support, he flounders about; and the more he struggles the more he entangles his shackles. The Spider, spying on the threshold of her abyss, lets him have his way. She does not run up the shrouds of the mast-work to seize the desperate prisoner; she waits until his bonds of threads, twisted backwards and forwards, make him fall on the web.

He falls; the other comes and flings herself upon her prostrate prey. The attack is not without danger. The Locust is demoralized rather than tied up; it is merely bits of broken thread that he is trailing from his legs. The bold assailant does not mind. Without troubling, like the Epeirae, to bury her capture under a paralysing winding-sheet, she feels it, to make sure of its quality, and then, regardless of kicks, inserts her fangs.

The bite is usually given at the lower end of a haunch: not that this place is more vulnerable than any other thin-skinned part, but probably because it has a better flavour. The different webs which I inspect to study the food in the larder show me, among other joints, various Flies and small Butterflies and carcasses of almost-untouched Locusts, all deprived of their hind-legs, or at least of one. Locusts' legs often dangle, emptied of their succulent contents, on the edges of the web, from the meat-hooks of the butcher's shop. In my urchin-days, days free from prejudices in regard to what one ate, I, like many others, was able to appreciate that dainty. It is the equivalent, on a very small scale, of the larger legs of the Crayfish.

The rigging-builder, therefore, to whom we have just thrown a Locust attacks the prey at the lower end of a thigh. The bite is a lingering one: once the Spider has planted her fangs, she does not let go. She drinks, she sips, she sucks. When this first point is drained, she passes on to others, to the second haunch in particular, until the prey becomes an empty hulk without losing its outline.

We have seen that Garden Spiders feed in a similar way, bleeding their venison and drinking it instead of eating it. At last, however, in the comfortable post-prandial hours, they take up the drained morsel, chew it, rechew it and reduce it to a shapeless ball. It is a dessert for the teeth to toy with. The Labyrinth Spider knows nothing of the diversions of the table; she flings the drained remnants out of her web, without chewing them. Although it lasts long, the meal is eaten in perfect safety. From the first bite, the Locust becomes a lifeless thing; the Spider's poison has settled him.

The labyrinth is greatly inferior, as a work of art, to that advanced geometrical contrivance, the Garden Spider's net; and, in spite of its ingenuity, it does not give a favourable notion of its constructor. It is hardly more than a shapeless scaffolding, run up anyhow. And yet, like the others, the builder of this slovenly edifice must have her own principles of beauty and accuracy. As it is, the prettily-latticed mouth of the crater makes us suspect this; the nest, the mother's usual masterpiece, will prove it to the full.

When laying-time is at hand, the Spider changes her residence; she abandons her web in excellent condition; she does not return to it. Whoso will can take possession of the house. The hour has come to found the family-establishment. But where? The Spider knows right well; I am in the dark. Mornings are spent in fruitless searches. In vain I ransack the bushes that carry the webs: I never find aught that realizes my hopes.

I learn the secret at last. I chance upon a web which, though deserted, is not yet dilapidated, proving that it has been but lately quitted. Instead of hunting in the brushwood whereon it rests, let us inspect the neighbourhood, to a distance of a few paces. If these contain a low, thick cluster, the nest is there, hidden from the eye. It carries an authentic certificate of its origin, for the mother invariably occupies it.

By this method of investigation, far from the labyrinth-trap, I become the owner of as many nests as are needed to satisfy my curiosity. They do not by a long way come up to my idea of the maternal talent. They are clumsy bundles of dead leaves, roughly drawn together with silk threads. Under this rude covering is a pouch of fine texture containing the egg-casket, all in very bad condition, because of the inevitable tears incurred in its extrication from the brushwood. No, I shall not be able to judge of the artist's capacity by these rags and tatters.

The insect, in its buildings, has its own architectural rules, rules as unchangeable as anatomical peculiarities. Each group builds according to the same set of principles, conforming to the laws of a very elementary system of aesthetics; but often circumstances beyond the architect's control—the space at her disposal, the unevenness of the site, the nature of the material and other accidental causes—interfere with the worker's plans and disturb the structure. Then virtual regularity is translated into actual chaos; order degenerates into disorder.

We might discover an interesting subject of research in the type adopted by each species when the work is accomplished without hindrances. The Banded Epeira weaves the wallet of her eggs in the open, on a slim branch that does not get in her way; and her work is a superbly artistic jar. The Silky Epeira also has all the elbow-room she needs; and her paraboloid is not without elegance. Can the Labyrinth Spider, that other spinstress of accomplished merit, be ignorant of the precepts of beauty when the time comes for her to weave a tent for her offspring? As yet, what I have seen of her work is but an unsightly bundle. Is that all she can do?

I look for better things if circumstances favour her. Toiling in the midst of a dense thicket, among a tangle of dead leaves and twigs, she may well produce a very inaccurate piece of work; but compel her to labour when free from all impediment: she will then—I am convinced of it beforehand—apply her talents without constraint and show herself an adept in the building of graceful nests.

As laying-time approaches, towards the middle of August, I instal half-a-dozen Labyrinth Spiders in large wire-gauze cages, each standing in an earthen pan filled with sand. A sprig of thyme, planted in the centre, will furnish supports for the structure, together with the trellis-work of the top and sides. There is no other furniture, no dead leaves, which would spoil the shape of the nest if the mother were minded to employ them as a covering. By way of provision, Locusts, every day. They are readily accepted, provided they be tender and not too large.

The experiment works perfectly. August is hardly over before I am in possession of six nests, magnificent in shape and of a dazzling whiteness. The latitude of the workshop has enabled the spinstress to follow the inspiration of her instinct without serious obstacles; and the result is a masterpiece of symmetry and elegance, if we allow for a few angularities demanded by the suspension-points.

It is an oval of exquisite white muslin, a diaphanous abode wherein the mother must make a long stay to watch over the brood.

The size is nearly that of a Hen's egg. The cabin is open at either end. The front-entrance broadens into a gallery; the back-entrance tapers into a funnel-neck. I fail to see the object of this neck. As for the opening in front, which is wider, this is, beyond a doubt, a victualling-door. I see the Spider, at intervals, standing here on the look-out for the Locust, whom she consumes outside, taking care not to soil the spotless sanctuary with corpses.

The structure of the nest is not without a certain similarity to that of the home occupied during the hunting-season. The passage at the back represents the funnel-neck, that ran almost down to the ground and afforded an outlet for flight in case of grave danger. The one in front, expanding into a mouth kept wide open by cords stretched backwards and forwards, recalls the yawning gulf into which the victims used to fall. Every part of the old dwelling is repeated: even the labyrinth, though this, it is true, is on a much smaller scale. In front of the bell-shaped mouth is a tangle of threads wherein the passers-by are caught. Each species, in this way, possesses a primary architectural model which is followed as a whole, in spite of altered conditions. The animal knows its trade thoroughly, but it does not know and will never know aught else, being incapable of originality.

Now this palace of silk, when all is said, is nothing more than a guard-house. Behind the soft, milky opalescence of the wall glimmers the egg-tabernacle, with its form vaguely suggesting the star of some order of knighthood. It is a large pocket, of a splendid dead-white, isolated on every side by radiating pillars which keep it motionless in the centre of the tapestry. These pillars are about ten in number and are slender in the middle, expanding at one end into a conical capital and at the other into a base of the same shape. They face one another and mark the position of the vaulted corridors which allow free movement in every direction around the central chamber. The mother walks gravely to and fro under the arches of her cloisters, she stops first here, then there; she makes a lengthy auscultation of the egg-wallet; she listens to all that happens inside the satin wrapper. To disturb her would be barbarous.

For a closer examination, let us use the dilapidated nests which we brought from the fields. Apart from its pillars, the egg-pocket is an inverted conoid, reminding us of the work of the Silky Epeira. Its material is rather stout; my pincers, pulling at it, do not tear it without difficulty. Inside the bag there is nothing but an extremely fine, white wadding and, lastly, the eggs, numbering about a hundred and comparatively large, for they measure a millimetre and a half [1]. They are very pale amber-yellow beads, which do not stick together and which roll freely as soon as I remove the swan's-down shroud. Let us put everything into a glass-tube to study the hatching.

We will now retrace our steps a little. When laying-time comes, the mother forsakes her dwelling, her crater into which her falling victims dropped, her labyrinth in which the flight of the Midges was cut short; she leaves intact the apparatus that enabled her to live at her ease. Thoughtful of her natural duties, she goes to found another establishment at a distance. Why at a distance?

She has still a few long months to live and she needs nourishment. Were it not better, then, to lodge the eggs in the immediate neighbourhood of the present home and to continue her hunting with the excellent snare at her disposal? The watching of the nest and the easy acquisition of provender would go hand in hand. The Spider is of another opinion; and I suspect the reason.

The sheet-net and the labyrinth that surmounts it are objects visible from afar, owing to their whiteness and the height whereat they are placed. Their scintillation in the sun, in frequented paths, attracts Mosquitoes and Butterflies, like the lamps in our rooms and the fowler's looking-glass. Whoso comes to look at the bright thing too closely dies the victim of his curiosity. There is nothing better for playing upon the folly of the passer-by, but also nothing more dangerous to the safety of the family.

Harpies will not fail to come running at this signal, showing up against the green; guided by the position of the web, they will assuredly find the precious purse; and a strange grub, feasting on a hundred new-laid eggs, will ruin the establishment. I do not know these enemies, not having sufficient materials at my disposal for a register of the parasites; but, from indications gathered elsewhere, I suspect them.

The Banded Epeira, trusting to the strength of her stuff, fixes her nest in the sight of all, hangs it on the brushwood, taking no

---

[1] .059 inch—*Translator's Note*.

precautions whatever to hide it. And a bad business it proves for her. Her jar provides me with an Ichneumon[1] possessed of the inoculating larding-pin: a *Cryptus* who, as a grub, had fed on Spiders' eggs. Nothing but empty shells was left inside the central keg; the germs were completely exterminated. There are other Ichneumon-flies, moreover, addicted to robbing Spiders' nests; a basket of fresh eggs is their offspring's regular food.

Like any other, the Labyrinth Spider dreads the scoundrelly advent of the pickwallet; she provides for it and, to shield herself against it as far as possible, chooses a hiding-place outside her dwelling, far removed from the tell-tale web. When she feels her ovaries ripen, she shifts her quarters; she goes off at night to explore the neighbourhood and seek a less dangerous refuge. The points selected are, by preference, the low brambles dragging along the ground, keeping their dense verdure during the winter and crammed with dead leaves from the oaks hard by. Rosemary-tufts, which gain in thickness what they lose in height on the unfostering rock, suit her particularly. This is where I usually find her nest, not without long seeking, so well is it hidden.

So far, there is no departure from current usage. As the world is full of creatures on the prowl for tender mouthfuls, every mother has her apprehensions; she also has her natural wisdom, which advises her to establish her family in secret places. Very few neglect this precaution; each, in her own manner, conceals the eggs she lays.

In the case of the Labyrinth Spider, the protection of the brood is complicated by another condition. In the vast majority of instances, the eggs, once lodged in a favourable spot, are abandoned to themselves, left to the chances of good or ill fortune. The Spider of the brushwood, on the contrary, endowed with greater maternal devotion, has, like the Crab Spider, to mount guard over hers until they hatch.

With a few threads and some small leaves joined together, the Crab Spider builds, above her lofty nest, a rudimentary watch-tower where she stays permanently, greatly emaciated, flattened into

---

[1] The Ichneumon-flies are very small insects which carry long ovipositors, wherewith they lay their eggs in the eggs of other insects and also, more especially, in caterpillars. Their parasitic larvae live and develop at the expense of the egg or grub attacked, which degenerates in consequence.— *Translator's Note*.

a sort of wrinkled shell through the emptying of her ovaries and the total absence of food. And this mere shred, hardly more than a skin that persists in living without eating, stoutly defends her egg-sack, shows fight at the approach of any tramp. She does not make up her mind to die until the little ones are gone.

The Labyrinth Spider is better treated. After laying her eggs, so far from becoming thin, she preserves an excellent appearance and a round belly. Moreover, she does not lose her appetite and is always prepared to bleed a Locust. She therefore requires a dwelling with a hunting-box close to the eggs watched over. We know this dwelling, built in strict accordance with artistic canons under the shelter of my cages.

Remember the magnificent oval guard-room, running into a vestibule at either end; the egg-chamber slung in the centre and isolated on every side by half a score of pillars; the front-hall expanding into a wide mouth and surmounted by a network of taut threads forming a trap. The semi-transparency of the walls allows us to see the Spider engaged in her household affairs. Her cloister of vaulted passages enables her to proceed to any point of the star-shaped pouch containing the eggs. Indefatigable in her rounds, she stops here and there; she fondly feels the satin, listens to the secrets of the wallet. If I shake the net at any point with a straw, she quickly runs up to enquire what is happening. Will this vigilance frighten off the Ichneumon and other lovers of omelettes? Perhaps so. But, though this danger be averted, others will come when the mother is no longer there.

Her attentive watch does not make her overlook her meals. One of the Locusts whereof I renew the supply at intervals in the cages is caught in the cords of the great entrance-hall. The Spider arrives hurriedly, snatches the giddy-pate and disjoints his shanks, which she empties of their contents, the best part of the insect. The remainder of the carcass is afterwards drained more or less, according to her appetite at the time. The meal is taken outside the guard-room, on the threshold, never indoors.

These are not capricious mouthfuls, serving to beguile the boredom of the watch for a brief while; they are substantial repasts, which require several sittings. Such an appetite astonishes me, after I have seen the Crab Spider, that no less ardent watcher, refuse the

Bees whom I give her and allow herself to die of inanition. Can this other mother have so great a need as that to eat? Yes, certainly she has; and for an imperative reason.

At the beginning of her work, she spent a large amount of silk, perhaps all that her reserves contained; for the double dwelling—for herself and for her offspring—is a huge edifice, exceedingly costly in materials; and yet, for nearly another month, I see her adding layer upon layer both to the wall of the large cabin and to that of the central chamber, so much so that the texture, which at first was translucent gauze, becomes opaque satin. The walls never seem thick enough; the Spider is always working at them. To satisfy this lavish expenditure, she must incessantly, by means of feeding, fill her silk-glands as and when she empties them by spinning. Food is the means whereby she keeps the inexhaustible factory going.

A month passes and, about the middle of September, the little ones hatch, but without leaving their tabernacle, where they are to spend the winter packed in soft wadding. The mother continues to watch and spin, lessening her activity from day to day. She recruits herself with a Locust at longer intervals; she sometimes scorns those whom I myself entangle in her trap. This increasing abstemiousness, a sign of decrepitude, slackens and at last stops the work of the spinnerets.

For four or five weeks longer, the mother never ceases her leisurely inspection-rounds, happy at hearing the new-born Spiders swarming in the wallet. At length, when October ends, she clutches her offspring's nursery and dies withered. She has done all that maternal devotion can do; the special providence of tiny animals will do the rest. When spring comes, the youngsters will emerge from their snug habitation, disperse all over the neighbourhood by the expedient of the floating thread and weave their first attempts at a labyrinth on the tufts of thyme.

Accurate in structure and neat in silk-work though they be, the nests of the caged captives do not tell us everything; we must go back to what happens in the fields, with their complicated conditions. Towards the end of December, I again set out in search, aided by all my youthful collaborators. We inspect the stunted rosemaries along the edge of a path sheltered by a rocky, wooded slope; we lift the branches that spread over the ground. Our zeal is rewarded with success. In a couple of hours, I am the owner of some nests.

Pitiful pieces of work are they, injured beyond recognition by the assaults of the weather! It needs the eyes of faith to see in these ruins the equivalent of the edifices built inside my cages. Fastened to the creeping branch, the unsightly bundle lies on the sand heaped up by the rains. Oak-leaves, roughly joined by a few threads, wrap it all round. One of these leaves, larger than the others, roofs it in and serves as a scaffolding for the whole of the ceiling. If we did not see the silky remnants of the two vestibules projecting and feel a certain resistance when separating the parts of the bundle, we might take the thing for a casual accumulation, the work of the rain and the wind.

Let us examine our find and look more closely into its shapelessness. Here is the large room, the maternal cabin, which rips as the coating of leaves is removed; here are the circular galleries of the guard-room; here are the central chamber and its pillars, all in a fabric of immaculate white. The dirt from the damp ground has not penetrated to this dwelling protected by its wrapper of dead leaves.

Now open the habitation of the offspring. What is this? To my utter astonishment, the contents of the chamber are a kernel of earthy matters, as though the muddy rain-water had been allowed to soak through. Put aside that idea, says the satin wall, which itself is perfectly clean inside. It is most certainly the mother's doing, a deliberate piece of work, executed with minute care. The grains of sand are stuck together with a cement of silk; and the whole resists the pressure of the fingers.

If we continue to unshell the kernel, we find, below this mineral layer, a last silken tunic that forms a globe around the brood. No sooner do we tear this final covering than the frightened little ones run away and scatter with an agility that is singular at this cold and torpid season.

To sum up, when working in the natural state, the Labyrinth Spider builds around the eggs, between two sheets of satin, a wall composed of a great deal of sand and a little silk. To stop the Ichneumon's probe and the teeth of the other ravagers, the best thing that occurred to her was this hoarding which combines the hardness of flint with the softness of muslin.

This means of defence seems to be pretty frequent among Spiders. Our own big House Spider, *Tegenaria domestica*, encloses her eggs in a globule strengthened with a rind of silk and of crumbly wreckage from the mortar of the walls. Other species, living in the open under stones, work in the same way. They wrap their eggs in a mineral shell held together with silk. The same fears have inspired the same protective methods.

Then how comes it that, of the five mothers reared in my cages, not one has had recourse to the clay rampart? After all, sand abounded: the pans in which the wire-gauze covers stood were full of it. On the other hand, under normal conditions, I have often come across nests without any mineral casing. These incomplete nests were placed at some height from the ground, in the thick of the brushwood; the others, on the contrary, those supplied with a coating of sand, lay on the ground.

The method of the work explains these differences. The concrete of our buildings is obtained by the simultaneous manipulation of gravel and mortar. In the same way, the Spider mixes the cement of the silk with the grains of sand; the spinnerets never cease working, while the legs fling under the adhesive spray the solid materials collected in the immediate neighbourhood. The operation would be impossible if, after cementing each grain of sand, it were necessary to stop the work of the spinnerets and go to a distance to fetch further stony elements. Those materials have to be right under her legs; otherwise the Spider does without and continues her work just the same.

In my cages, the sand is too far off. To obtain it, the Spider would have to leave the top of the dome, where the nest is being built on its trellis-work support; she would have to come down some nine inches. The worker refuses to take this trouble, which, if repeated in the case of each grain, would make the action of the spinnerets too irksome. She also refuses to do so when, for reasons which I have not fathomed, the site chosen is some way up in the tuft of rosemary. But, when the nest touches the ground, the clay rampart is never missing.

Are we to see in this fact proof of an instinct capable of modification, either making for decadence and gradually neglecting what was the ancestors' safeguard, or making for progress and advancing, hesitatingly, towards perfection in the mason's art? No inference is permissible in either direction. The Labyrinth Spider has simply taught us that instinct possesses resources which are employed or left latent according to the conditions of the moment. Place sand under her legs and the spinstress will knead concrete; refuse her that sand, or put it out of her reach, and the Spider will remain a simple silk-worker, always ready, however, to turn mason under favourable conditions. The aggregate of things that come within the observer's scope proves that it were mad to expect from her any further innovations, such as would utterly change her methods of manufacture and cause her, for instance, to abandon her cabin, with its two entrance-halls and its star-like tabernacle, in favour of the Banded Epeira's pear-shaped gourd.

CHAPTER XVI

THE CLOTHO SPIDER

SHE is named Durand's Clotho (*Clotho Durandi*, LATR.), in memory of him who first called attention to this particular Spider. To enter on eternity under the safe-conduct of a diminutive animal which saves us from speedy oblivion under the mallows and rockets is no contemptible advantage. Most men disappear without leaving an echo to repeat their name; they lie buried in forgetfulness, the worst of graves.

Others, among the naturalists, benefit by the designation given to this or that object in life's treasure-house: it is the skiff wherein they keep afloat for a brief while. A patch of lichen on the bark of an old tree, a blade of grass, a puny beastie: any one of these hands down a man's name to posterity as effectively as a new comet. For all its abuses, this manner of honouring the departed is eminently respectable. If we would carve an epitaph of some duration, what could we find better than a Beetle's wing-case, a Snail's shell or a Spider's web? Granite is worth none of them. Entrusted to the hard stone, an inscription becomes obliterated; entrusted to a Butterfly's wing, it is indestructible. 'Durand,' therefore, by all means.

But why drag in 'Clotho'? Is it the whim of a nomenclator, at a loss for words to denote the ever-swelling tide of beasts that require cataloguing? Not entirely. A mythological name came to his mind, one which sounded well and which, moreover, was not out of place in designating a spinstress. The Clotho of antiquity is the youngest of the three Fates; she holds the distaff whence our destinies are spun, a distaff wound with plenty of rough flocks, just a few shreds of silk and, very rarely, a thin strand of gold.

Prettily shaped and clad, as far as a Spider can be, the Clotho of the naturalists is, above all, a highly talented spinstress; and this is the reason why she is called after the distaff-bearing deity of the infernal regions. It is a pity that the analogy extends no further. The mythological Clotho, niggardly with her silk and lavish with her coarse flocks, spins us a harsh existence; the eight-legged Clotho uses naught but exquisite silk. She works for herself; the other works for us, who are hardly worth the trouble.

Would we make her acquaintance? On the rocky slopes in the oliveland, scorched and blistered by the sun, turn over the flat stones, those of a fair size; search, above all, the piles which the shepherds set up for a seat whence to watch the sheep browsing amongst the lavender below. Do not be too easily disheartened: the Clotho is rare; not every spot suits her. If fortune smile at last upon our perseverance, we shall see, clinging to the lower surface of the stone which we have lifted, an edifice of a weather-beaten aspect, shaped like an over-turned cupola and about the size of half a tangerine orange. The outside is encrusted or hung with small shells, particles of earth and, especially, dried insects.

The edge of the cupola is scalloped into a dozen angular lobes, the points of which spread and are fixed to the stone. In between these straps is the same number of spacious inverted arches. The whole represents the Ishmaelite's camel-hair tent, but upside down. A flat roof, stretched between the straps, closes the top of the dwelling.

Then where is the entrance? All the arches of the edge open upon the roof; not one leads to the interior. The eye seeks in vain; there is nothing to point to a passage between the inside and the outside. Yet the owner of the house must go out from time to time, were it only in search of food; on returning from her expedition, she must go in again. How does she make her exits and her entrances? A straw will tell us the secret.

Pass it over the threshold of the various arches. Everywhere, the searching straw encounters resistance; everywhere, it finds the place rigorously closed. But one of the scallops, differing in no wise from the others in appearance, if cleverly coaxed, opens at the edge into two lips and stands slightly ajar. This is the door, which at once shuts again of its own elasticity. Nor is this all: the Spider, when she returns home, often bolts herself in, that is to say, she joins and fastens the two leaves of the door with a little silk.

The Mason Mygale is no safer in her burrow, with its lid undistinguishable from the soil and moving on a hinge, than is the Clotho in her tent, which is inviolable by any enemy ignorant of the device. The Clotho, when in danger, runs quickly home; she opens the chink with a touch of her claw, enters and disappears. The door closes of itself and is supplied, in case of need, with a lock consisting of a few threads. No burglar, led astray by the multiplicity of arches, one and all alike, will ever discover how the fugitive vanished so suddenly.

While the Clotho displays a more simple ingenuity as regards her defensive machinery, she is incomparably ahead of the Mygale in the matter of domestic comfort. Let us open her cabin. What luxury! We are taught how a Sybarite of old was unable to rest, owing to the presence of a crumpled rose-leaf in his bed. The Clotho is quite as fastidious. Her couch is more delicate than swan's-down and whiter than the fleece of the clouds where brood the summer storms. It is the ideal blanket. Above is a canopy or tester of equal softness. Between the two nestles the Spider, short-legged, clad in sombre garments, with five yellow favours on her back.

Rest in this exquisite retreat demands perfect stability, especially on gusty days, when sharp draughts penetrate beneath the stone. This condition is admirably fulfilled. Take a careful look at the habitation. The arches that gird the roof with a balustrade and bear the weight of the edifice are fixed to the slab by their extremities. Moreover, from each point of contact, there issues a cluster of diverging threads that creep along the stone and cling to it throughout their length, which spreads afar. I have measured some fully nine inches long. These are so many cables; they represent the ropes and pegs that hold the Arab's tent in position. With such supports as these, so numerous and so methodically arranged, the hammock cannot be torn from its bearings save by the intervention of brutal methods with which the Spider need not concern herself, so seldom do they occur.

Another detail attracts our attention: whereas the interior of the house is exquisitely clean, the outside is covered with dirt, bits of earth, chips of rotten wood, little pieces of gravel. Often there are worse things still: the exterior of the tent becomes a charnel-house. Here, hung up or embedded, are the dry carcasses of Opatra, Asidae and other Tenebrionidae[1] that favour underrock shelters; segments of Iuli[2], bleached by the sun; shells of Pupae[3], common among the stones; and, lastly, Snail-shells, selected from among the smallest.

These relics are obviously, for the most part, table-leavings, broken victuals. Unversed in the trapper's art, the Clotho courses her game and lives upon the vagrants who wander from one stone to another. Whoso ventures under the slab at night is strangled by the hostess; and the dried-up carcass, instead of being flung to a distance, is hung to the silken wall, as though the Spider wished to make a bogey-house of her home. But this cannot be her aim. To act like the ogre who hangs his victims from the castle battlements is the worst way to disarm suspicion in the passers-by whom you are lying in wait to capture.

There are other reasons which increase our doubts. The shells hung up are most often empty; but there are also some occupied by the Snail, alive and untouched. What can the Clotho do with a *Pupa cinerea*, a *Pupa quadridens* and other narrow spirals wherein the animal retreats to an inaccessible depth? The Spider is incapable of breaking the calcareous shell or of getting at the hermit through the opening. Then why should she collect those prizes, whose slimy flesh is probably not to her taste? We begin to suspect a simple question of ballast and balance. The House Spider, or *Tegenaria domestica*, prevents her web, spun in a corner of the wall, from losing its shape at the least breath of air, by loading it with crumbling plaster and allowing tiny fragments of mortar to accumulate. Are we face to face with a similar process? Let us try experiment, which is preferable to any amount of conjecture.

To rear the Clotho is not an arduous undertaking; we are not obliged to take the heavy flagstone, on which the dwelling is built, away with us. A very simple operation suffices. I loosen the fastenings with my pocket-knife. The Spider has such stay-at-home ways that she very rarely makes off. Besides, I use the utmost

---

[1] One of the largest families of Beetles, darkish in colour and shunning the light.—*Translator's Note*.

[2] The Iulus is one of the family of Myriapods, which includes Centipedes, etc.—*Translator's Note*.

[3] A species of Land-snail.—*Translator's Note*.

discretion in my rape of the house. And so I carry away the building, together with its owner, in a paper bag.

The flat stones, which are too heavy to move and which would occupy too much room upon my table, are replaced either by deal disks, which once formed part of cheese-boxes, or by round pieces of cardboard. I arrange each silken hammock under one of these by itself, fastening the angular projections, one by one, with strips of gummed paper. The whole stands on three short pillars and gives a very fair imitation of the underrock shelter in the form of a small dolmen. Throughout this operation, if you are careful to avoid shocks and jolts, the Spider remains indoors. Finally, each apparatus is placed under a wire-gauze, bell-shaped cage, which stands in a dish filled with sand.

We can have an answer by the next morning. If, among the cabins swung from the ceilings of the deal or cardboard dolmens, there be one that is all dilapidated, that was seriously knocked out of shape at the time of removal, the Spider abandons it during the night and instals herself elsewhere, sometimes even on the trellis-work of the wire cage.

The new tent, the work of a few hours, attains hardly the diameter of a two-franc piece. It is built, however, on the same principles as the old manor-house and consists of two thin sheets laid one above the other, the upper one flat and forming a tester, the lower curved and pocket-shaped. The texture is extremely delicate: the least trifle would deform it, to the detriment of the available space, which is already much reduced and only just sufficient for the recluse.

Well, what has the Spider done to keep the gossamer stretched, to steady it and to make it retain its greatest capacity? Exactly what our static treatises would advise her to do: she has ballasted her structure, she has done her best to lower its centre of gravity. From the convex surface of the pocket hang long chaplets of grains of sand strung together with slender silken cords. To these sandy stalactites, which form a bushy beard, are added a few heavy lumps hung separately and lower down, at the end of a thread. The whole is a piece of ballast-work, an apparatus for ensuring equilibrium and tension.

The present edifice, hastily constructed in the space of a night, is the frail rough sketch of what the home will afterwards become. Successive layers will be added to it; and the partition-wall will grow into a thick blanket capable of partly retaining, by its own weight, the requisite curve and capacity. The Spider now abandons the stalactites of sand, which were used to keep the original pocket stretched, and confines herself to dumping down on her abode any more or less heavy object, mainly corpses of insects, because she need not look for these and finds them ready to hand after each meal. They are weights, not trophies; they take the place of materials that must otherwise be collected from a distance and hoisted to the top. In this way, a breastwork is obtained that strengthens and steadies the house. Additional equilibrium is often supplied by tiny shells and other objects hanging a long way down.

What would happen if one robbed an old dwelling, long since completed, of its outer covering? In case of such a disaster, would the Spider go back to the sandy stalactites, as a ready means of restoring stability? This is easily ascertained. In my hamlets under wire, I select a fair-sized cabin. I strip the exterior, carefully removing any foreign body. The silk reappears in its original whiteness. The tent looks magnificent, but seems to me too limp.

This is also the Spider's opinion. She sets to work, next evening, to put things right. And how? Once more with hanging strings of sand. In a few nights, the silk bag bristles with a long, thick beard of stalactites, a curious piece of work, excellently adapted to maintain the web in an unvaried curve. Even so are the cables of a suspension-bridge steadied by the weight of the superstructure.

Later, as the Spider goes on feeding, the remains of the victuals are embedded in the wall, the sand is shaken and gradually drops away and the home resumes its charnel-house appearance. This brings us to the same conclusion as before: the Clotho knows her statics; by means of additional weights, she is able to lower the centre of gravity and thus to give her dwelling the proper equilibrium and capacity.

Now what does she do in her softly-wadded home? Nothing, that I know of. With a full stomach, her legs luxuriously stretched over the downy carpet, she does nothing, thinks of nothing; she listens to the sound of earth revolving on its axis. It is not sleep, still

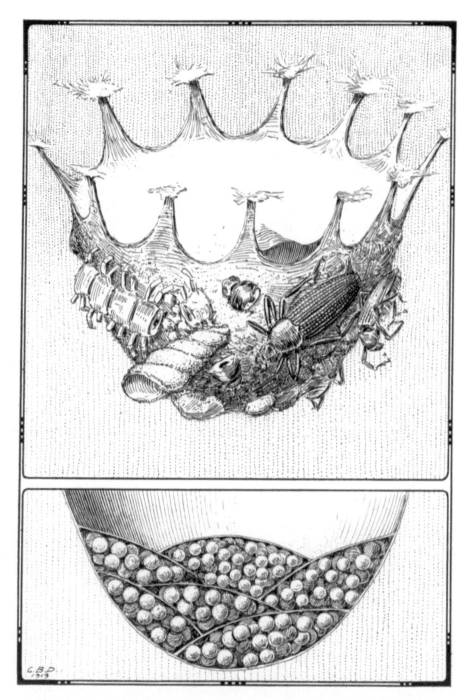

NEXT OF THE CLOTHO SPIDER, COVERED WITH THE REMAINS OF THE FEAST (UPPER)

CAPSULES OF EGGS IN THE BOTTOM OF CLOTHO'S NEXT (LOWER)

less is it waking; it is a middle state where naught prevails save a dreamy consciousness of well-being. We ourselves, when comfortably in bed, enjoy, just before we fall asleep, a few moments of bliss, the prelude to cessation of thought and its train of worries; and those moments are among the sweetest in our lives. The Clotho seems to know similar moments and to make the most of them.

If I push open the door of the cabin, invariably I find the Spider lying motionless, as though in endless meditation. It needs the teasing of a straw to rouse her from her apathy. It needs the prick of hunger to bring her out of doors; and, as she is extremely temperate, her appearances outside are few and far between. During three years of assiduous observation, in the privacy of my study, I have not once seen her explore the domain of the wire cage by day. Not until a late hour at night does she venture forth in quest of victuals; and it is hardly feasible to follow her on her excursions.

Patience once enabled me to find her, at ten o'clock in the evening, taking the air on the flat roof of her house, where she was doubtless waiting for the game to pass. Startled by the light of my candle, the lover of darkness at once returned indoors, refusing to reveal any of her secrets. Only, next day, there was one more corpse hanging from the wall of the cabin, a proof that the chase was successfully resumed after my departure.

The Clotho, who is not only nocturnal, but also excessively shy, conceals her habits from us; she shows us her works, those precious historical documents, but hides her actions, especially the laying, which I estimate approximately to take place in October. The sum total of the eggs is divided into five or six small, flat, lentiform pockets, which, taken together, occupy the greater part of the maternal home. These capsules have each their own partition-wall of superb white satin, but they are so closely soldered, both together and to the floor of the house, that it is impossible to part them without tearing them, impossible, therefore, to obtain them separately. The eggs in all amount to about a hundred.

The mother sits upon the heap of pockets with the same devotion as a brooding hen. Maternity has not withered her. Although decreased in bulk, she retains an excellent look of health; her round belly and her well-stretched skin tell us from the first that her part is not yet wholly played.

The hatching takes place early. November has not arrived before the pockets contain the young: wee things clad in black, with five yellow specks, exactly like their elders. The new-born do not leave their respective nurseries. Packed close together, they spend the whole of the wintry season there, while the mother, squatting on the pile of cells, watches over the general safety, without knowing her family other than by the gentle trepidations felt through the partitions of the tiny chambers. The Labyrinth Spider has shown us how she maintains a permanent sitting for two months in her guard-room, to defend, in case of need, the brood which she will never see. The Clotho does the same during eight months, thus earning the right to set eyes for a little while on her family trotting around her in the main cabin and to assist at the final exodus, the great journey undertaken at the end of a thread.

When the summer heat arrives, in June, the young ones, probably aided by their mother, pierce the walls of their cells, leave the maternal tent, of which they know the secret outlet well, take the air on the threshold for a few hours and then fly away, carried to some distance by a funicular aeroplane, the first product of their spinning-mill.

The elder Clotho remains behind, careless of this emigration which leaves her alone. She is far from being faded indeed, she looks younger than ever. Her fresh colour, her robust appearance suggest great length of life, capable of producing a second family. On this subject I have but one document, a pretty far-reaching one, however. There were a few mothers whose actions I had the patience to watch, despite the wearisome minutiae of the rearing and the slowness of the result. These abandoned their dwellings after the departure of their young; and each went to weave a new one for herself on the wire net-work of the cage.

They were rough-and-ready summaries, the work of a night. Two hangings, one above the other, the upper one flat, the lower concave and ballasted with stalactites of grains of sand, formed the new home, which, strengthened daily by fresh layers, promised to become similar to the old one. Why does the Spider desert her former mansion, which is in no way dilapidated—far from it—and still exceedingly serviceable, as far as one can judge? Unless I am mistaken, I think I have an inkling of the reason.

The old cabin, comfortably wadded though it be, possesses serious disadvantages: it is littered with the ruins of the children's nurseries. These ruins are so close-welded to the rest of the home that my forceps cannot extract them without difficulty; and to remove them would be an exhausting business for the Clotho and possibly beyond her strength. It is a case of the resistance of Gordian knots, which not even the very spinstress who fastened them is capable of untying. The encumbering litter, therefore, will remain.

If the Spider were to stay alone, the reduction of space, when all is said, would hardly matter to her: she wants so little room, merely enough to move in! Besides, when you have spent seven or eight months in the cramping presence of those bedchambers, what can be the reason of a sudden need for greater space? I see but one: the Spider requires a roomy habitation, not for herself—she is satisfied with the smallest den—but for a second family. Where is she to place the pockets of eggs, if the ruins of the previous laying remain in the way? A new brood requires a new home. That, no doubt, is why, feeling that her ovaries are not yet dried up, the Spider shifts her quarters and founds a new establishment.

The facts observed are confined to this change of dwelling. I regret that other interests and the difficulties attendant upon a long upbringing did not allow me to pursue the question and definitely to settle the matter of the repeated layings and the longevity of the Clotho, as I did in that of the Lycosa.

Before taking leave of this Spider, let us glance at a curious problem which has already been set by the Lycosa's offspring. When carried for seven months on the mother's back, they keep in training as agile gymnasts without taking any nourishment. It is a familiar exercise for them, after a fall, which frequently occurs, to scramble up a leg of their mount and nimbly to resume their place in the saddle. They expend energy without receiving any material sustenance.

The sons of the Clotho, the Labyrinth Spider and many others confront us with the same riddle: they move, yet do not eat. At any period of the nursery stage, even in the heart of winter, on the bleak days of January, I tear the pockets of the one and the tabernacle of the other, expecting to find the swarm of youngsters lying in a state of complete inertia, numbed by the cold and by lack of food. Well, the result is quite different. The instant their cells are broken open, the anchorites run out and flee in every direction as nimbly as at the best moments of their normal liberty. It is marvellous to see them scampering about. No brood of Partridges, stumbled upon by a Dog, scatters more promptly.

Chicks, while still no more than tiny balls of yellow fluff, hasten up at the mother's call and scurry towards the plate of rice. Habit has made us indifferent to the spectacle of those pretty little animal machines, which work so nimbly and with such precision; we pay no attention, so simple does it all appear to us. Science examines and looks at things differently. She says to herself:

'Nothing is made with nothing. The chick feeds itself; it consumes or rather it assimilates and turns the food into heat, which is converted into energy.'

Were any one to tell us of a chick which, for seven or eight months on end, kept itself in condition for running, always fit, always brisk, without taking the least beakful of nourishment from the day when it left the egg, we could find no words strong enough to express our incredulity. Now this paradox of activity maintained without the stay of food is realized by the Clotho Spider and others.

I believe I have made it sufficiently clear that the young Lycosae take no food as long as they remain with their mother. Strictly speaking, doubt is just admissible, for observation is needs dumb as to what may happen earlier or later within the mysteries of the burrow. It seems possible that the repleted mother may there disgorge to her family a mite of the contents of her crop. To this suggestion the Clotho undertakes to make reply.

Like the Lycosa, she lives with her family; but the Clotho is separated from them by the walls of the cells in which the little ones are hermetically enclosed. In this condition, the transmission of solid nourishment becomes impossible. Should any one entertain a theory of nutritive humours cast up by the mother and filtering through the partitions at which the prisoners might come and drink, the Labyrinth Spider would at once dispel the idea. She dies a few weeks after her young are hatched; and the children, still locked in their satin bed-chamber for the best part of the year, are none the less active.

Can it be that they derive sustenance from the silken wrapper? Do they eat their house? The supposition is not absurd, for we have seen the Epeirae, before beginning a new web, swallow the ruins of the old. But the explanation cannot be accepted, as we learn from the Lycosa, whose family boasts no silky screen. In short, it is certain that the young, of whatever species, take absolutely no nourishment.

Lastly, we wonder whether they may possess within themselves reserves that come from the egg, fatty or other matters the gradual combustion of which would be transformed into mechanical force. If the expenditure of energy were of but short duration, a few hours or a few days, we could gladly welcome this idea of a motor viaticum, the attribute of every creature born into the world. The chick possesses it in a high degree: it is steady on its legs, it moves for a little while with the sole aid of the food wherewith the egg furnishes it; but soon, if the stomach is not kept supplied, the centre of energy becomes extinct and the bird dies. How would the chick fare if it were expected, for seven or eight months without stopping, to stand on its feet, to run about, to flee in the face of danger? Where would it stow the necessary reserves for such an amount of work?

The little Spider, in her turn, is a minute particle of no size at all. Where could she store enough fuel to keep up mobility during so long a period? The imagination shrinks in dismay before the thought of an atom endowed with inexhaustible motive oils.

We must needs, therefore, appeal to the immaterial, in particular to heat-rays coming from the outside and converted into movement by the organism. This is nutrition of energy reduced to its simplest expression: the motive heat, instead of being extracted from the food, is utilized direct, as supplied by the sun, which is the seat of all life. Inert matter has disconcerting secrets, as witness radium; living matter has secrets of its own, which are more wonderful still. Nothing tells us that science will not one day turn the suspicion suggested by the Spider into an established truth and a fundamental theory of physiology.

APPENDIX

THE GEOMETRY OF THE EPEIRA'S WEB

I FIND myself confronted with a subject which is not only highly interesting, but somewhat difficult: not that the subject is obscure; but it presupposes in the reader a certain knowledge of geometry: a strong meat too often neglected. I am not addressing geometricians, who are generally indifferent to questions of instinct, nor entomological collectors, who, as such, take no interest in mathematical theorems; I write for any one with sufficient intelligence to enjoy the lessons which the insect teaches.

What am I to do? To suppress this chapter were to leave out the most remarkable instance of Spider industry; to treat it as it should be treated, that is to say, with the whole armoury of scientific formulae, would be out of place in these modest pages. Let us take a middle course, avoiding both abstruse truths and complete ignorance.

Let us direct our attention to the nets of the Epeirae, preferably to those of the Silky Epeira and the Banded Epeira, so plentiful in the autumn, in my part of the country, and so remarkable for their bulk. We shall first observe that the radii are equally spaced; the angles formed by each consecutive pair are of perceptibly equal value; and this in spite of their number, which in the case of the Silky Epeira exceeds two score. We know by what strange means the Spider attains her ends and divides the area wherein the web is to be warped into a large number of equal sectors, a number which is almost invariable in the work of each species. An operation without method, governed, one might imagine, by an irresponsible whim, results in a beautiful rose-window worthy of our compasses.

We shall also notice that, in each sector, the various chords, the elements of the spiral windings, are parallel to one another and gradually draw closer together as they near the centre. With the two radiating lines that frame them they form obtuse angles on one side and acute angles on the other; and these angles remain constant in the same sector, because the chords are parallel.

There is more than this: these same angles, the obtuse as well as the acute, do not alter in value, from one sector to another, at any rate so far as the conscientious eye can judge. Taken as a whole, therefore, the rope-latticed edifice consists of a series of cross-bars intersecting the several radiating lines obliquely at angles of equal value.

By this characteristic we recognize the 'logarithmic spiral.' Geometricians give this name to the curve which intersects obliquely, at angles of unvarying value, all the straight lines or 'radii vectores' radiating from a centre called the 'Pole.' The Epeira's construction, therefore, is a series of chords joining the intersections of a logarithmic spiral with a series of radii. It would become merged in this spiral if the number of radii were infinite, for this would reduce the length of the rectilinear elements indefinitely and change this polygonal line into a curve.

To suggest an explanation why this spiral has so greatly exercised the meditations of science, let us confine ourselves for the present to a few statements of which the reader will find the proof in any treatise on higher geometry.

The logarithmic spiral describes an endless number of circuits around its pole, to which it constantly draws nearer without ever being able to reach it. This central point is indefinitely inaccessible at each approaching turn. It is obvious that this property is beyond our sensory scope. Even with the help of the best philosophical instruments, our sight could not follow its interminable windings and would soon abandon the attempt to divide the invisible. It is a volute to which the brain conceives no limits. The trained mind, alone, more discerning than our retina, sees clearly that which defies the perceptive faculties of the eye.

The Epeira complies to the best of her ability with this law of the endless volute. The spiral revolutions come closer together as they approach the pole. At a given distance, they stop abruptly; but, at this point, the auxiliary spiral, which is not destroyed in the central region, takes up the thread; and we see it, not without some surprise, draw nearer to the pole in ever-narrowing and scarcely perceptible circles. There is not, of course, absolute mathematical accuracy, but a very close approximation to that accuracy. The Epeira winds nearer and nearer round her pole, so far as her equipment, which, like our own, is defective, will allow her. One would believe her to be thoroughly versed in the laws of the spiral.

I will continue to set forth, without explanations, some of the properties of this curious curve. Picture a flexible thread wound round a logarithmic spiral. If we then unwind it, keeping it taut the while, its free extremity will describe a spiral similar at all points to the original. The curve will merely have changed places.

Jacques Bernouilli[1], to whom geometry owes this magnificent theorem, had engraved on his tomb, as one of his proudest titles to fame, the generating spiral and its double, begotten of the unwinding of the thread. An inscription proclaimed, '*Eadem mutata resurgo*: I rise again like unto myself.' Geometry would find it difficult to better this splendid flight of fancy towards the great problem of the hereafter.

There is another geometrical epitaph no less famous. Cicero, when quaestor in Sicily, searching for the tomb of Archimedes amid the thorns and brambles that cover us with oblivion, recognized it, among the ruins, by the geometrical figure engraved upon the stone: the cylinder circumscribing the sphere. Archimedes, in fact, was the first to know the approximate relation of circumference to diameter; from it he deduced the perimeter and surface of the circle, as well as the surface and volume of the sphere. He showed that the surface and volume of the last-named equal two-thirds of the surface and volume of the circumscribing cylinder. Disdaining all pompous inscription, the learned Syracusan honoured himself with his theorem as his sole epitaph. The geometrical figure proclaimed the individual's name as plainly as would any alphabetical characters.

To have done with this part of our subject, here is another property of the logarithmic spiral. Roll the curve along an indefinite straight line. Its pole will become displaced while still

---

[1] Jacques Bernouilli (1654-1705), professor of mathematics at the University of Basel from 1687 to the year of his death. He improved the differential calculus, solved the isoperimetrical problem and discovered the properties of the logarithmic spiral.—*Translator's Note*.

keeping on one straight line. The endless scroll leads to rectilinear progression; the perpetually varied begets uniformity.

Now is this logarithmic spiral, with its curious properties, merely a conception of the geometers, combining number and extent, at will, so as to imagine a tenebrous abyss wherein to practise their analytical methods afterwards? Is it a mere dream in the night of the intricate, an abstract riddle flung out for our understanding to browse upon?

No, it is a reality in the service of life, a method of construction frequently employed in animal architecture. The Mollusc, in particular, never rolls the winding ramp of the shell without reference to the scientific curve. The first-born of the species knew it and put it into practice; it was as perfect in the dawn of creation as it can be to-day.

Let us study, in this connection, the Ammonites, those venerable relics of what was once the highest expression of living things, at the time when the solid land was taking shape from the oceanic ooze. Cut and polished length-wise, the fossil shows a magnificent logarithmic spiral, the general pattern of the dwelling which was a pearl palace, with numerous chambers traversed by a siphuncular corridor.

To this day, the last representative of the Cephalopoda with partitioned shells, the Nautilus of the Southern Seas, remains faithful to the ancient design; it has not improved upon its distant predecessors. It has altered the position of the siphuncle, has placed it in the centre instead of leaving it on the back, but it still whirls its spiral logarithmically as did the Ammonites in the earliest ages of the world's existence.

And let us not run away with the idea that these princes of the Mollusc tribe have a monopoly of the scientific curve. In the stagnant waters of our grassy ditches, the flat shells, the humble Planorbes, sometimes no bigger than a duckweed, vie with the Ammonite and the Nautilus in matters of higher geometry. At least one of them, *Planorbis vortex*, for example, is a marvel of logarithmic whorls.

In the long-shaped shells, the structure becomes more complex, though remaining subject to the same fundamental laws. I have before my eyes some species of the genus Terebra, from New

Caledonia. They are extremely tapering cones, attaining almost nine inches in length. Their surface is smooth and quite plain, without any of the usual ornaments, such as furrows, knots or strings of pearls. The spiral edifice is superb, graced with its own simplicity alone. I count a score of whorls which gradually decrease until they vanish in the delicate point. They are edged with a fine groove.

I take a pencil and draw a rough generating line to this cone; and, relying merely on the evidence of my eyes, which are more or less practised in geometric measurements, I find that the spiral groove intersects this generating line at an angle of unvarying value.

The consequence of this result is easily deduced. If projected on a plane perpendicular to the axis of the shell, the generating lines of the cone would become radii; and the groove which winds upwards from the base to the apex would be converted into a plane curve which, meeting those radii at an unvarying angle, would be neither more nor less than a logarithmic spiral. Conversely, the groove of the shell may be considered as the projection of this spiral on a conic surface.

Better still. Let us imagine a plane perpendicular to the aids of the shell and passing through its summit. Let us imagine, moreover, a thread wound along the spiral groove. Let us unroll the thread, holding it taut as we do so. Its extremity will not leave the plane and will describe a logarithmic spiral within it. It is, in a more complicated degree, a variant of Bernouilli's *'Eadem mutata resurgo:'* the logarithmic conic curve becomes a logarithmic plane curve.

A similar geometry is found in the other shells with elongated cones, Turritellae, Spindle-shells, Cerithia, as well as in the shells with flattened cones, Trochidae, Turbines. The spherical shells, those whirled into a volute, are no exception to this rule. All, down to the common Snail-shell, are constructed according to logarithmic laws. The famous spiral of the geometers is the general plan followed by the Mollusc rolling its stone sheath.

Where do these glairy creatures pick up this science? We are told that the Mollusc derives from the Worm. One day, the Worm, rendered frisky by the sun, emancipated itself, brandished its tail and twisted it into a corkscrew for sheer glee. There and then the plan of the future spiral shell was discovered.

This is what is taught quite seriously, in these days, as the very last word in scientific progress. It remains to be seen up to what point the explanation is acceptable. The Spider, for her part, will have none of it. Unrelated to the appendix-lacking, corkscrew-twirling Worm, she is nevertheless familiar with the logarithmic spiral. From the celebrated curve she obtains merely a sort of framework; but, elementary though this framework be, it clearly marks the ideal edifice. The Epeira works on the same principles as the Mollusc of the convoluted shell.

The Mollusc has years wherein to construct its spiral and it uses the utmost finish in the whirling process. The Epeira, to spread her net, has but an hour's sitting at the most, wherefore the speed at which she works compels her to rest content with a simpler production. She shortens the task by confining herself to a skeleton of the curve which the other describes to perfection.

The Epeira, therefore, is versed in the geometric secrets of the Ammonite and the *Nautilus pompilus*; she uses, in a simpler form, the logarithmic line dear to the Snail. What guides her? There is no appeal here to a wriggle of some kind, as in the case of the Worm that ambitiously aspires to become a Mollusc. The animal must needs carry within itself a virtual diagram of its spiral. Accident, however fruitful in surprises we may presume it to be, can never have taught it the higher geometry wherein our own intelligence at once goes astray, without a strict preliminary training.

Are we to recognize a mere effect of organic structure in the Epeira's art? We readily think of the legs, which, endowed with a very varying power of extension, might serve as compasses. More or less bent, more or less outstretched, they would mechanically determine the angle whereat the spiral shall intersect the radius; they would maintain the parallel of the chords in each sector.

Certain objections arise to affirm that, in this instance, the tool is not the sole regulator of the work. Were the arrangement of the thread determined by the length of the legs, we should find the spiral volutes separated more widely from one another in proportion to the greater length of implement in the spinstress. We see this in the Banded Epeira and the Silky Epeira. The first has longer limbs and spaces her cross-threads more liberally than does the second, whose legs are shorter.

But we must not rely too much on this rule, say others. The Angular Epeira, the Paletinted Epeira and the Cross Spider, all three more or less short-limbed, rival the Banded Epeira in the spacing of their lime-snares. The last two even dispose them with greater intervening distances.

We recognize in another respect that the organization of the animal does not imply an immutable type of work. Before beginning the sticky spiral, the Epeirae first spin an auxiliary intended to strengthen the stays. This spiral, formed of plain, non-glutinous thread, starts from the centre and winds in rapidly-widening circles to the circumference. It is merely a temporary construction, whereof naught but the central part survives when the Spider has set its limy meshes. The second spiral, the essential part of the snare, proceeds, on the contrary, in serried coils from the circumference to the centre and is composed entirely of viscous cross-threads.

Here we have, following one after the other merely by a sudden alteration of the machine, two volutes of an entirely different order as regards direction, the number of whorls and intersection. Both of them are logarithmic spirals. I see no mechanism of the legs, be they long or short, that can account for this alteration.

Can it then be a premeditated design on the part of the Epeira? Can there be calculation, measurement of angles, gauging of the parallel by means of the eye or otherwise? I am inclined to think that there is none of all this, or at least nothing but an innate propensity, whose effects the animal is no more able to control than the flower is able to control the arrangement of its verticils. The Epeira practises higher geometry without knowing or caring. The thing works of itself and takes its impetus from an instinct imposed upon creation from the start.

The stone thrown by the hand returns to earth describing a certain curve; the dead leaf torn and wafted away by a breath of wind makes its journey from the tree to the ground with a similar curve. On neither the one side nor the other is there any action by the moving body to regulate the fall; nevertheless, the descent takes place according to a scientific trajectory, the 'parabola,' of which the section of a cone by a plane furnished the prototype to the geometer's speculations. A figure, which was at first but a tentative glimpse, becomes a reality by the fall of a pebble out of the vertical.

The same speculations take up the parabola once more, imagine it rolling on an indefinite straight line and ask what course does the focus of this curve follow. The answer comes: The focus of the parabola describes a 'catenary,' a line very simple in shape, but endowed with an algebraic symbol that has to resort to a kind of cabalistic number at variance with any sort of numeration, so much so that the unit refuses to express it, however much we subdivide the unit. It is called the number e. Its value is represented by the following series carried out ad infinitum:

$$e = 1 + \frac{1}{1} + \frac{1}{1.2} + \frac{1}{1.2.3} + \frac{1}{1.2.3.4} + \frac{1}{1.2.3.4.5} + \text{etc.}$$

If the reader had the patience to work out the few initial terms of this series, which has no limit, because the series of natural numerals itself has none, he would find:

$$e = 2.7182818...$$

With this weird number are we now stationed within the strictly defined realm of the imagination? Not at all: the catenary appears actually every time that weight and flexibility act in concert. The name is given to the curve formed by a chain suspended by two of its points which are not placed on a vertical line. It is the shape taken by a flexible cord when held at each end and relaxed; it is the line that governs the shape of a sail bellying in the wind; it is the curve of the nanny-goat's milk-bag when she returns from filling her trailing udder. And all this answers to the number e.

What a quantity of abstruse science for a bit of string! Let us not be surprised. A pellet of shot swinging at the end of a thread, a drop of dew trickling down a straw, a splash of water rippling under the kisses of the air, a mere trifle, after all, requires a titanic scaffolding when we wish to examine it with the eye of calculation. We need the club of Hercules to crush a fly.

Our methods of mathematical investigation are certainly ingenious; we cannot too much admire the mighty brains that have invented them; but how slow and laborious they appear when compared with the smallest actualities! Will it never be given to us to probe reality in a simpler fashion? Will our intelligence be able one day to dispense with the heavy arsenal of formulae? Why not?

Here we have the abracadabric number e reappearing, inscribed on a Spider's thread. Let us examine, on a misty morning, the meshwork that has been constructed during the night. Owing to their hygrometrical nature, the sticky threads are laden with tiny drops, and, bending under the burden, have become so many catenaries, so many chaplets of limpid gems, graceful chaplets arranged in exquisite order and following the curve of a swing. If the sun pierce the mist, the whole lights up with iridescent fires and becomes a resplendent cluster of diamonds. The number e is in its glory.

Geometry, that is to say, the science of harmony in space, presides over everything. We find it in the arrangement of the scales of a fir-cone, as in the arrangement of an Epeira's limy web; we find it in the spiral of a Snail-shell, in the chaplet of a Spider's thread, as in the orbit of a planet; it is everywhere, as perfect in the world of atoms as in the world of immensities.

And this universal geometry tells us of an Universal Geometrician, whose divine compass has measured all things. I prefer that, as an explanation of the logarithmic curve of the Ammonite and the Epeira, to the Worm screwing up the tip of its tail. It may not perhaps be in accordance with latter-day teaching, but it takes a loftier flight.